ELEMENTS OF POLICE SUPERVISION

Second Edition

GLENCOE CRIMINAL JUSTICE SERIES

Allen/Simonsen: **Corrections in America: An Introduction (Second Edition)**
Bloomquist: **Marijuana: The Second Trip**
Brandstatter/Hyman: **Fundamentals of Law Enforcement**
Coffey/Eldefonso: **Process and Impact of Justice**
Eldefonso: **Issues in Corrections**
Eldefonso: **Readings in Criminal Justice**
Eldefonso/Coffey: **Process and Impact of the Juvenile Justice System**
Eldefonso/Hartinger: **Control, Treatment, and Rehabilitation of Juvenile Offenders**
Engel/DeGreen/Rebo: **The Justice Game: A Simulation**
Gourley: **Effective Municipal Police Organization**
LeGrande: **The Basic Processes of Criminal Justice**
Lentini: **Vice and Narcotics Control**
Kamon/Hunt/Fleming: **Juvenile Law and Procedures in California (Revised Edition)**
Melnicoe/Mennig: **Elements of Police Supervision (Second Edition)**
Nelson: **Preliminary Investigation and Police Reporting: A Complete Guide to Police Written Communication**
Pursley: **Introduction to Criminal Justice**
Radelet: **The Police and the Community (Second Edition)**
Radelet/Reed: **The Police and the Community: Studies**
Roberts/Bristow: **An Introduction to Modern Police Firearms**
Simonsen/Gordon: **Juvenile Justice in America**
Waddington: **Arrest, Search, and Seizure**
Waddington: **Criminal Evidence**
Wicks/Platt: **Drug Abuse: A Criminal Justice Primer**

General Editor:
G. DOUGLAS GOURLEY

Professor and Chairman
Department of Criminal Justice
California State University at Los Angeles
Los Angeles, California

ELEMENTS OF POLICE SUPERVISION
SECOND EDITION

WILLIAM B. MELNICOE
Chairman, Department of Criminal Justice
and Forensic Science
California State University
Sacramento, California

JAN C. MENNIG
Culver City, California

GLENCOE PUBLISHING CO., INC.
Encino, California

Collier Macmillan Publishers
London

Copyright © 1978 Glencoe Publishing Co., Inc.
Earlier edition copyright © 1969 by Glencoe Press,
a division of Macmillan Publishing Co., Inc.

Printed in the United States of America

All rights reserved. No part of this book may be reproduced or transmitted in any form or by any means, electronic or mechanical, including photocopying, recording, or by any information storage and retrieval system, without permission in writing from the Publisher.

Glencoe Publishing Co., Inc.
17337 Ventura Boulevard
Encino, California 91316
Collier Macmillan Canada, Ltd.

Library of Congress Catalog Card Number: 76-4022

1 2 3 4 5 6 7 8 9   80 79 78 77

ISBN 0-02-476000-5

CONTENTS

	PREFACE	vii
1.	INTRODUCTION TO SUPERVISION	1
2.	THE POLICE ENVIRONMENT	7
3.	THE SUPERVISOR'S ROLE IN MANAGEMENT	16
4.	THE SELECTION OF SUPERVISORS	39
5.	THE PSYCHOLOGICAL ASPECTS OF SUPERVISION	50
6.	MOTIVATION	63
7.	LEADERSHIP	71
8.	MORALE	92
9.	EMPLOYEE GROUPS AND ORGANIZATIONS	103
10.	DISCIPLINE	125
11.	COMMUNICATION	148
12.	COUNSELING AND INTERVIEWING	168
13.	COMPLAINTS AND GRIEVANCES	177
14.	DECISION MAKING AND PLANNING	188
15.	WORK PLANNING	205

16. PERFORMANCE APPRAISAL	213
17. WOMEN AND THE SUPERVISORY ROLE	221
18. THE SUPERVISORY TRAINING FUNCTION	228

APPENDIX

A.	Example of a General Order: Complaints of Police Personnel Misconduct Any City, U.S.A. Police Department	256
B.	Complaint Form Oakland, California Police Department	260
C.	Personnel Evaluation Form	261
D.	Example of a General Order: Permanent Employment Status—Achievement Any City, U.S.A. Police Department	264
E.	Supervisory Training Course State of New York Municipal Police Training Council	270
F.	Qualities of Leadership—A Supervisor's Rating Scale California Program for Peace Officers' Training	275
G.	A Form for Rating Supervisors California Program for Peace Officers' Training	278
H.	Self-Analysis Questionnaire for Supervisors California Program for Peace Officers' Training	280
I.	Field Training Guide Berkeley, California Police Department	285
J.	Courses and Programs California Commission on Peace Officers' Standards and Training Specifications	300

INDEX	311

Preface

This second edition of *Elements of Police Supervision* represents the most contemporary thinking in the field and covers every advance in supervisory techniques, with a special emphasis on the unique problems of police officers. The authors have brought together in one volume what they believe to be the best practices of both government and industry adapted to the special problems of the criminal justice system. As such, this text will be equally applicable to any and all agencies within criminal jurisprudence.

This book was made possible largely because of the efforts of many individuals and the organizations they represent, who contributed much of the material. The authors are particularly grateful to the training units of the Ford Motor Company, Lockheed Aircraft Corporation, and North American Aviation, who gave generously of time and supervisory training materials. We also acknowledge the generosity of Mr. Robert D. Gray, Director of the Industrial Relations Center, California Institute of Technology, and his institution, for permission to adapt and use material from the publication, *Supervision of Engineers*. The authors also wish to express thanks to the Department of the California Highway Patrol for permission to adapt and use material from *Field Management Series,* Vol. 7: *Supervision and Leadership,* and to the Department of the Army for use of Pamphlet 5-2 and other relevant army publications on staff, supervisory, and training matters. Particular thanks is also due the Peace Officers' Training Unit, California State Department of Education, for generously contributing much material to this volume and granting permission to reprint material which was published in *Supervisory Personnel Development* by William Melnicoe and John Peper.

Additionally, the authors wish to acknowledge Mr. Douglas E. Millham and Mr. William E. Eastman for their assistance in the preparation of the second edition of this book.

This definitive text will prove helpful to any man or woman who aspires to be a police supervisor and to any supervisor who sincerely desires to improve his or her leadership abilities.

1 Introduction to Supervision

"Good supervision is one of the cornerstones upon which successful police operations rest. There is no police agency, regardless of its size or scope of operations, that can function at maximum efficiency unless it maintains high quality supervisory control over all its personnel."[1]

Professionalization of the police service is one of the most pressing problems facing American society today, because only highly professional police officers are able to cope with the complicated problems and social pressures that accompany the struggle for civil rights. Too often the police are caught between opposing factions of society, and no matter what action is taken by law enforcement officers, it will seem wrong to one faction or another. The police officer is squarely confronted with one of the major dilemmas of the day—damned if action is taken and damned if it isn't.

Today's officer is also confronted with a steadily increasing crime rate, juvenile delinquency, and impossible traffic flow problems. Solutions to these problems are being sought, but they will not be found in the immediate future. However, the seeking of solutions will be a major problem for the police as well as for other agencies of government. When solutions are found, proper implementation of policies and procedures will rest with the supervisors of the agencies. Thus, the supervisory role in police service will be increasingly significant.

The Importance of Good Supervision

The public's image of a police agency is determined largely by the relationship between officers and citizens.[2] Both good police service and misconduct by police officers are noted by the public and reflect favorably or unfavorably on the entire force. Good supervision, then, is essential in all agencies concerned with the administration of justice and is, in fact, their obligation to the public.

The development of efficient supervisory personnel is one of the major problems of the police administrator. The need for such personnel is strongly indicated by: (1) the rapidly expanding and growing urban community, (2) the shifts in population patterns, (3) the pressure of an ever-increasing crime rate, (4) the high personnel turnover in police agencies, and (5) the lack of experienced personnel. A member of police management today is not as personally and professionally close to former partners and peers as was previously possible. The new supervisor must develop an entirely new outlook toward job responsibilities, and now must suddenly adapt to a new concept of accomplishing work through others instead of doing it personally.

Even the experienced supervisor is confronted with problems that simply did not exist in the "good old days." Frequently, being not really equipped to solve such problems and preferring to ignore them, the supervisor merely hopes they will go away eventually. (They won't—they simply get worse when left unattended.)

Because of the great need for today's police officers to find solutions to a wide range of difficult problems, it is imperative that all police agencies be directed and controlled by trained professionals. Generally, nonprofessionals have proved to be blundering and incompetent.

Even the professional police officer can fail with insufficient training in the principles of organization and administration and their applications. Well-trained professionals also have failed because they placed insufficient emphasis on supervision. A police chief may be able to establish high standards for the selection and training of subordinates, but without close, high-quality supervision, even benefits such as high salaries, paid overtime, liberal vacation and pension plans, etc., will not prevent demoralization of these personnel. All of the essential principles of management, including supervision, must have appropriate weight. To be effective, supervision must be continuous and constructive.[3]

Every modern scientific device should be utilized in the selection of police supervisors. Ordinary trial-and-error methods of examination may fail to uncover personality handicaps that render an officer unfit for performing supervisory functions. Those charged with the responsibility of selecting police supervisors should realize the fallacy of believing that any good officer can become an excellent supervisor without professional training for the task. A lack of cooperation on the part of police personnel can usually be traced to unsuitable supervision. *Supervisors who prove to be misfits should be eliminated without delay.*

The Job of the Police Supervisor

The job of the police supervisor is most exacting and includes many different responsibilities, among which are the following:

1. The police supervisor must make certain that all subordinate personnel have the requisite qualifications for the positions they occupy and are placed where they can best serve. Rules governing lines of authority must be enforced. The efforts of all individuals and groups must be coordinated, and friction between individuals and groups must be reduced in order to maintain morale at the highest possible level. The supervisor must watch for personality defects in subordinates and must note the effects of time and of social, economic, and political change upon them.

2. The supervisor must enforce the observance of rules and regulations, general and special orders, and departmental policies; this enforcement must be humane and rational.

3. The supervisor must constantly search for flaws in the structure of the department. Weaknesses in operational plans must be detected and faulty procedures corrected. The reasons for failures in any phase of the department's operation must be discovered and thoroughly analyzed.

4. The supervisor must see that all subordinates are striving to achieve departmental objectives.

5. The supervisor must also take responsibility for training, planning, counseling, and motivating subordinate personnel.

All of these concepts will be discussed in depth in subsequent chapters. *Webster's New World Dictionary* defines "supervision" as "being supervised; direction; management" and a "supervisor" as "a person who supervises; superintendent; manager; director." In this text, the designation "supervisor" shall apply to all persons, male or female, with the rank of sergeant or above, with the exception of the director, commissioner, sheriff, chief, or superintendent of police. The supervisor's rank and responsibilities depend largely upon the size, organization, and structure of the individual agency. In some agencies, the head of the organization may have supervisory responsibilities, while in others a sergeant may have no such responsibilities.

As referred to in this text, "police service" includes all persons actively engaged in law enforcement, crime prevention, and the repression of criminal activities.

Supervisory Development

Supervisors can make or break any organization. On a nationwide basis, comparatively few police organizations have really done anything in the field of supervisory training. Because the supervisor "builds" subordinate personnel, the success of the entire department depends largely on those who occupy this position. Supervisors can build good relationships with subordinates and the public or can develop antagonism and mistrust.

A good example of the feeling of private industry regarding supervisory training is reflected in this statement of policy from the company manual of the Ford Motor Company:

> Qualified replacements will be developed for all key positions. In the selection of a new employee, emphasis will be placed on the applicant's potential. Once hired, he will be given every opportunity for development. Work assignments should be challenging. He should be given additional training where appropriate. Performance should be reviewed with him frequently and his accomplishments should be rewarded with promotions, salary increases, and, where appropriate, supplemental compensation. To the extent practicable, an individual development plan will be formulated for each key employee. The plan will provide for broad and progressively more responsible work experience to the extent permitted by the employee's potential rate of progress.[4]

Training programs for nonsupervisory personnel have been heavily emphasized in the police service. However, little if any consideration was given to the training of supervisors until California and New York, through their respective Commissions on Peace Officers' Standards and Training, took the lead. Outlines of the required courses for new supervisors in New York and California are included in the appendix. The authors anticipate that the recommendations made in a report by the President's Commission on Law Enforcement and Administration of Justice will add impetus to proper selection and training of police supervisory personnel in the future.[5] It is interesting to note that one of the recommendations contained in another report of the Commission is that all persons promoted to supervisory positions be required to hold at least a baccalaureate degree from an accredited college or university.[6] The Commission considers this a first step toward the eventual requirement of a bachelor's degree for all police personnel.

The Supervisor and the Subordinate

In any organization, public or private, supervision basically involves a one-to-one relationship. It follows, then, that supervision must be permissive enough to allow subordinates to develop to their full capacity.[7] This does not by any means imply that the word *boss* has become obsolete. To some of the proponents of the human relations approach to management, the concept of bossing is no longer in vogue. But a realistic consideration of the supervisor-subordinate relationship indicates that a certain amount of bossing is normal and inevitable.

Most people regard the relationship between themselves and their boss as one of the most important in their lives. It vitally affects their performance on the job, as well as their interpersonal relationships with others off the job. The officer who comes home from work, angry about what was considered an unjust "chewing out" by a sergeant, is quite likely to take out frustrations on family members. If unable to relieve frustrations at home and still angry when returning to work for the next tour of duty, the officer's tensions are likely to "snowball." Under such circumstances, one cannot possibly perform at a maximum level of efficiency. The officer's approach both to job and to personal life is drastically altered, and both job and personal life suffer as a result.

The supervisor-subordinate relationship is conditioned also by the climate of the organization within which it operates; this climate is, in turn, conditioned by many opposing forces. The supervisor must be responsive to the wishes of police management and must enforce those policies and procedures that are set forth by the management. At the same time, the management in the police agency must be responsive to the governing power of the community and, ultimately, to the people. Unfortunately for the present-day police administrator, there is no agreement among different segments of the people as to what is expected or wanted from their police agency. The police administrator is subject to many severe pressures from outside the organization to follow particular courses of action, and at the same time is subjected to pressures just as severe to do that which is diametrically opposed to these actions. Under such pressures, it is not surprising that police policies are subject to day-to-day changes which result in confusion to supervisors and their subordinates.

The Tools of the Supervisor

To best be effective, it is essential that the supervisor be given the tools necessary to carry out the supervisory assignment. Effectiveness can only be achieved through a well-organized training program built around the needs of the department and the individual supervisor. Some of those major needs are: (1) the development of a department manual that will provide a clear understanding of the purpose of the department's policies and procedures, and (2) the development of skills and techniques for the supervisor in the areas of human relations, organization and administration, applied psychology, public relations, work-performance appraisal, teaching skills, departmental rules and regulations, etc. While this list is extremely limited, it should be sufficient to indicate that broader educational and training programs are badly needed to provide the insights necessary to anyone who is to fill a supervisory position, whether it is at the top or the bottom level of the organization.

In constructing this text, the authors have attempted to cover those areas of supervision that are common to all police agencies. Some of the principles stated here may not be directly applicable in all departments, but it is hoped that they will provide the basis for solutions to local problems. It is not possible to develop a text covering all problems that may arise in the field of police supervision. And the study of supervision and leadership will *not* provide rules, formulas, or methods that fit all groups in any given situation. Books and courses on supervision are useless unless the principles and concepts are put into practice and properly applied by the supervisor. The supervisor alone can do this, and unless he or she is willing to transform the principles into techniques and to practice them, this person will be wasting time studying.

DISCUSSION QUESTIONS

1. What are some of the considerations that make good supervision essential to the police operation?
2. Why may trial-and-error be an unsatisfactory method of selecting supervisors?
3. What are some of a supervisor's responsibilities?
4. Why are broad education and training programs needed for supervisory personnel?

NOTES

1. William B. Melnicoe and John P. Peper, *Supervisory Personnel Development* (Sacramento, Calif.: California State Department of Education, 1965), p. vii.
2. Ibid.
3. Ibid., p. 2.
4. Ibid., p. 4.
5. President's Commission on Law Enforcement and Administration of Justice, *The Challenge of Crime in a Free Society* (Washington, D.C.: Government Printing Office, 1967).
6. Ibid., p. 127.
7. John M. Pfiffner and Marshall Fels, *The Supervision of Personnel,* 3rd ed. (Englewood Cliffs, N.J.: Prentice-Hall, Inc., 1964), p. 3.

2 The Police Environment

In recent years the growth of most urban communities and many rural areas has accelerated at a tremendous rate, and with this dramatic expansion have come pressures and responsibilities upon law enforcement never previously experienced. In addition to growth, American society has produced a rapidly fluctuating economy, an employment outlook which is, at best, unstable and economically disastrous for many minorities, and a myriad of other dilemmas which affect the stability of our communities.

Inasmuch as the police officer is primarily involved with the resolution or adjudication of many of the resulting social problems, the supervisor must be intensely aware of the constantly changing social environment in which law enforcement functions. The supervisor occupies a critical position in this regard, for the continual information which he or she receives from subordinates on the changing nature of the community can provide decision makers with a more accurate basis for program planning, implementation, and evaluation.

Contemporary administrative philosophy is to utilize this resource and have the first-line supervisor function as an active participant in administrative decision making, thus making maximum use of the supervisor's input.

In this chapter, the environment of law enforcement will be discussed in both historical and contemporary context. Guidelines will be suggested for the best utilization of police resources, the development of departmental goals and objectives, and the introduction of a contemporary management system utilizing a team process.

The Historical Role

The function of policing is to control the behavior of individuals or groups acting against the safety of persons or property. By custom and religion, certain acts are labeled as wrongs against society. These acts are also considered antisocial in nature. Classes of crimes or offenses against the state have emerged from the codification of law and regulations. The passing of laws covering crimes and offenses is part of the behavior of a modern, complex society. Authoritative sources have suggested that much of the world's population is still policed in the same fashion as society has protected itself since the beginning of recorded history. In *Police Administration,* John P. Kenney states, "It is a folk system for policing which relies on control methods established by the family, the community or the tribal leaders or councils. The prevailing customs prescribe the system."[1]

Law enforcement history indicates that the police task has been one of considerable difficulty. In a society such as ours, with its congestion and anonymity, it is even more obvious that the police function means different things to different people, including the police. Many practitioners as well as writers in the fields of political science and public administration have called for reassessment and rethinking of the police function in American society.

Today, there are over forty thousand separate agencies in the United States for the enforcement of law at the federal, state, and local levels. These agencies are not evenly distributed among the three levels. There are approximately fifty law enforcement agencies at the federal level, two hundred at the state level, and the remaining number are distributed among the many counties, cities, villages, and towns across the nation. These local police agencies have held fast to their traditional authority and responsibility for maintaining public order. What has emerged is a vast array of decentralized organizations which comprise the bulk of law enforcement influence in the United States. To say that this represents an unwieldy problem is an understatement, as these agencies employ in excess of 420,000 full- and part-time law enforcement officers and civilians. At the local level, in excess of 300,000 officers are divided among the county and local police agencies.

Policing practices in the United States suggest that the police have a preventive role to perform in the course of their activities. Some local U.S. police organizations have adopted a preventive enforcement philosophy and believe that the police should apply their knowledge and efforts toward limiting the opportunity and curbing the desire to commit crime. This philosophy is hard to identify, measure, and evaluate by traditional standards. The time-honored peace-keeping function of the police, based on legislation, has been to preserve order within a community—a function which requires a variety of skills. The police also perform emergency services consistent with their skills and are available on a 24-hour-a-day basis. Considering the inventory of tasks required by police organizations, there appears to be an absence of well-developed policies that can guide police personnel in handling the variety of activities in which they are engaged. Until recently, there have been few efforts to utilize any formal planning process to develop policies and guidelines to assist police personnel in the exercise of their roles. The overall criminal justice system itself is composed of a number of subsystems which, at times, work at cross-purposes. Police administrators have an important policy-making responsibility consistent with their vital role in society. Meaningful goals and objectives must be established.

Law enforcement must be identified as a process and not a purpose. The authors believe that the primary role or goal of police in modern society is the maintenance of order, without loss of liberty. The criminal aspects of law enforcement have been overemphasized. The pure "cops-and-robbers" police role covers a relatively small portion of the job of maintaining order. Recognized police authorities have indicated that approximately 98% of current police activity involves social service assistance dedicated to the prevention of crime and to the development of an environment of stability and security. These activities have been identified by Kenney in his writings as noncoercive, whereas criminal, traffic, juvenile, and regulatory law enforcement has been labeled as coercive. We must recognize yet another major responsibility of the police—the protection of civil rights and personal liberties. When coupled with those functional areas previously identified, this responsibility should give initial direction for the organization, administration, and operation of a local police agency.

Organization and Structure

In its broadest sense, police administration includes all of the activities of the federal, state, and local governments related to the accomplishment of the police function. Historically, the Wickersham Commission, in the thirties, along with *Municipal Police Administration,* published by the International City Managers' Association in 1938, and O. W. Wilson's *Police Administration,* published in 1950, have emerged as the principal texts influencing police organization in the United States. The succeeding editions of *Municipal Police Administration* and Wilson's *Police Administration,* along with more recent texts, have continued to be the sources of many more recent police changes.

During the past decade, considerable progress has been made in police organization, personnel management, and operations. This progress has not been limited to urban departments alone. Many sheriff's departments have also modernized their activities in order to provide better levels of local protection. Unfortunately, this has not been effective enough, and it is evident that far more "miles per gallon" can be obtained by keeping pace with the technological advances of other disciplines. Some agencies have taken advantage of those resources while, unfortunately, others have not.

State and local police agencies have, through the years, viewed themselves as semimilitary organizations, even looking to the military for organizational and some operational guidelines. Military ranks were also adopted by the police units.

The basic policies or guidelines for the administration and operation of the police function were derived from a variety of sources, such as the United States Constitution, federal law, state constitutions and laws, and county and city charters, laws, and ordinances. Administrative guidelines were developed internally through the influence of legislative bodies and administrators. The various departmental guidelines were drawn up by the agencies performing the police function.

Throughout this process, a further integration of the behavioral sciences was indicated. The formulation of police policy must be developed from a systematic process in which important issues regarding operational behavior of people within the organization are identified, studied, and resolved. Police organizations must engage in this process and develop guidelines covering their various operations. They must also provide for a modification of guidelines based on court decisions, public pressures, and changes in legal and administrative decisions established at a higher level. Policy should have local legislative endorsement.

The Three Basic Organizational Models

An examination of the behavior of local police agencies indicates that three organizational models have generally emerged: the traditional plan, the Berkeley or Vollmer plan, and the modified plan. In the traditional plan, the street, or beat, police officer performs minimal street police tasks, with specialists completing any follow-up work required. The Berkeley or Vollmer plan involves use of the beat police officer as the heart of the operation, performing all police tasks required within his or her geographic area of responsibility. In the modified plan, the beat or street police officer performs a wide variety of tasks, which includes preliminary and follow-up investigations, with referral in those cases requiring special expertise.

Police departments which have adopted the traditional model usually have not had a very professional view toward police work and certainly have not maximized their resources in terms of coordinated efforts and internal human relations. These departments have not focused on common police objectives within the various functional units. It is obvious that overspecialization results from the use of this model.

The structure of the Berkeley or Vollmer model and the modified model is basically the same. In essence, these plans use a three-divisional configuration of investigation, field, and services, supervised by an office of the chief of police which contains staff activities of personnel and training, planning, and budgeting. This organizational model plan has been easy for the majority of police agencies in the United States to identify with, as most agencies are small, with one to forty personnel. Surveys indicate that most of the smaller police agencies are organized along this pattern. These departments are typically divided into three shifts, or watches, and are somewehat self-contained. The personnel on duty handle all situations requiring police attention. Follow-up investigation or case processing may involve the use of specialists, designated as investigators or detectives. The chief, in this type of setting, wears many hats and, in addition to performing administrative functions, may serve as a follow-up investigator. Fairly large departments have been formed using this model, with more internal levels in the hierarchy and far greater specialization of functions. The success of this model has not been as great as was expected. Consequently, public administrators, police officials, and academicians in the field have sought new and better ways to organize the police, borrowing from the advanced technologies of other disciplines.

The Colleague Model

Robert T. Golembiewski has introduced a structuralist application, which he describes in his book, *Organizing Men and Power*.[2] He identifies the organization as a large, complex social unit in which many social groups interact, and he maintains that a synthesis of these group interactions in a manner that reflects the expectations of both management and the workers is the desirable approach to organization. This approach combines features of both the traditional formal organization and the human relations informal organization. In application, the model discards the classical model plans and integrates the concepts of the human relations approach into a new theory of formal organization. This model, known as the colleague model, calls for organizing assigned sub-units that focus on program activities, with integration of essential sustaining or staff activities into each sub-unit. The colleague model is basically a team approach to organization. Relationships are not hierarchical in the traditional sense. The top level of the organization is composed of management personnel with particular expertise in a functional or program area, working together as a team to provide direction in achieving organizational goals and objectives. Sub-units of the top management units may be formed to accomplish specific functions in a program area. The program teams are responsible for accomplishing the tasks of the organization. The emphasis on programs requires that participating employees develop a new look at organization and demands a professional quality of employees at all levels. The teams require the expertise necessary to perform all essential activities of the organization. For the performance of basic field activities, an operational team response may consist of a number of generalist police officers for routine activities, and follow-up experts for the more involved types of case investigations and related tasks. The product is a self-contained unit capable of performing all essential police activities, assuring specific responsibility and accountability for performance. This concept is designed to involve all personnel in the department in the achievement of established goals and objectives.

In his book *Police Administration,* John P. Kenney deals with the role of police in modern society.[3] He has developed the Golembiewski colleague model into a viable police application. Kenney's contribution is the first major input toward breaking the logjam of much current police thinking.

MBO/MBR/TBO

Management by objectives, also referred to as management by results, is a concept only recently introduced in a way that can be practically applied to the criminal justice field. Supervisors who would seek increased efficiency and effectiveness of subordinates' decision-making capabilities, as well as their own, would do well to familiarize themselves with this concept. In simplest terms, management by objectives (MBO), or management by results (MBR), places responsibility on the supervisor or manager to describe a desirable "future state" for subordinates to attain. He or she sets high standards for subordinates and defines results which will not only utilize their best abilities but also enlarge them. The supervisor is more concerned with letting people achieve their objectives than with controlling their specific behavior. The supervisor's concern extends to defining results, training those who cannot behave in a way which will get results, and defining any limits to the means subordinates may use to get results. The supervisor measures performance in terms of results which were agreed upon prior to the period in which the measurement occurs.

TBO, or training by objectives, is an effort to obtain a more precise definition of training for purposes of problem solving and decision making. In this sense training is defined as "planned changing of behavior." Behavior is the specific activity of people which can be seen or measured, and it may change for a variety of reasons. The system may affect behavior, or supervision may affect behavior. When the requirement is a planned course of coaching, classes, or guided experience originated and designed for the specific purpose of changing behavior, it is a training problem. How can you determine when you have a training problem?

1. You have a problem when something has deviated from the norm.

2. You have eliminated the possibility that it is a systems problem.

3. You have found that supervisory control is breaking down and you obtain further evidence that control by the supervisor is impossible because the people involved simply can't behave in a suitable fashion. Possibly, they don't know how.

4. There have probably been changes in people occupying certain positions, or the people are the same but the job has changed in a way that requires new behavior which they cannot immediately adopt.

Training problems fitting these criteria often have generated a series of further decisions which must be made before suitable action can be taken. Can the problem be solved by a systems change, thereby averting the necessity for training? This alternative is based on the knowledge that training is often more time-consuming, more resistant to success, and more

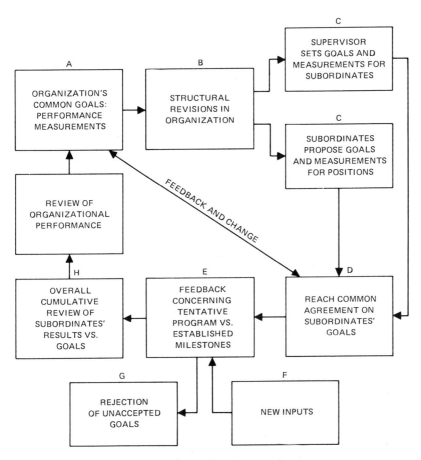

FIGURE 1. Management by Objectives/Results

costly than systems changes. It may be better to redesign a reporting form, making it easier to fill out, than to attempt to train the officers to fill out an extremely complex form. The likelihood of getting better results is in-

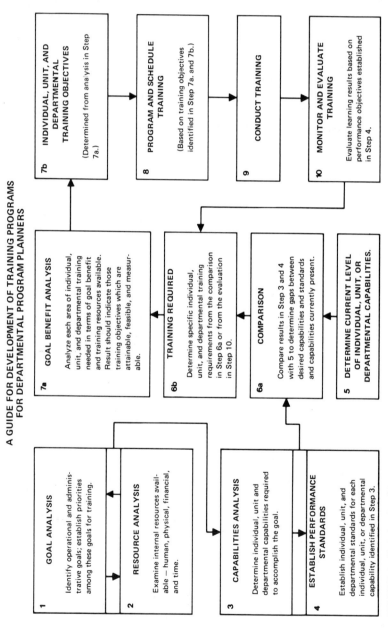

FIGURE 2. Training by Objectives

creased. The purpose is to get the information, not to harass police officers. The system may itself be a shaper of behavior if it presents stimuli which make the desired behavior more likely. It may also remove stimuli which encourage undesirable behavior. The flow charts included in Figures 1 and 2 diagram the steps which should be taken in a thorough application of MBO/TBO principles.

DISCUSSION QUESTIONS

1. In what ways have the traditional police roles and functions changed or expanded in recent years?
2. In terms of the several organizational models discussed, how does the role of the supervisor vary from one to another?
3. How is formal departmental policy developed, and what factors can influence its formulation?
4. What is meant by the term, "maximum utilization of resources"?
5. What is a team approach to decision making? How can this approach be expanded to include field operations? To include all aspects of administrative and operational police functions?
6. How may MBO/TBO concepts result in benefits to the department? To the supervisor? To the individual officer?

NOTES

1. John P. Kenney, *Police Administration* (Springfield, Ill.: Charles C. Thomas, 1975), p. 3.
2. Robert T. Golembiewski, *Organizing Men and Power: Patterns of Behavior and Line-Staff Models* (Chicago: Rand McNally & Co., 1967).
3. John P. Kenney, op. cit.

3 The Supervisor's Role in Management

The police supervisor is an integral part of the management team within the department and, as such, must be vitally concerned with the accomplishment of basic police purposes within the policy framework of the organization. Too many new supervisors think of themselves as peers in their relationships with subordinates and thus cannot perform effectively in their roles as supervisors. The supervisor must control and direct subordinates and must never confuse the police supervisory function with that of a union shop steward who constantly bucks management on behalf of labor. The supervisor must learn to think in terms of management in order to be effective.

The supervisor must play a dual role: representing department administration to subordinates and acting as the subordinates' representative to management. When supervision is either poor or overbalanced on the side of management, subordinates will soon feel that the department is not a good place to work. The supervisor must accurately communicate subordinates' problems and feelings to the administration because this information is essential to top-level decision making. The supervisor will be judged by the performance of subordinates, and so must recognize the necessity of looking out for their interests as a matter of self-preservation, if for no other reason.

The supervisor, standing between subordinates and the command group, is in an ideal position to prevent misunderstandings between the two groups. In this position, the supervisor should be prepared to interpret subordinates to the management and the management to subordinates.

Nearly all managers in all types of organizations have a common denominator—subordinates, those people who work under them. For this reason the presence or absence of subordinates is a useful factor in defining who is a manager. Table 1 shows four different sets of classifications which might be used to identify a management work force. Each set is interchangeable with the others when dealing with police organizations, although it must be recognized that the exact titles of police ranks may vary from one department to another, depending on size and geographic location. The executive group in one organization may be called either the top managers, the fourth level (if the fourth level is the highest level), or the chief and deputy chiefs in a police agency. There is always a degree of overlap within organizations, and it is often difficult to define sharply where middle management or supervision begins and ends within any given organization. For example, all of the deputy chiefs in a large police agency may not be included when the chief calls an executive meeting.

Even with these difficulties and apparent differences, it is possible to group managers using the concepts of classification and subordinates. In general, it can be said that supervisors usually have subordinates at the level at which day-to-day operations take place, whereas middle managers and second- and third-level personnel frequently are the bosses of other managers. Thus, having subordinates identifies a manager, and the kind of subordinates identifies the level of the manager.

I By Title	II By Position	III By Level	IV By Rank
Executives	Top	4th Level	Chief Deputy Chief
Managers	Middle	3rd Level 2nd Level	Captain Lieutenant
Supervisors	Bottom	1st Level	Sergeant

Table 1.

The world of the supervisor is constantly changing, yet some of the people within that world will resist change. No two days are the same to a supervisor in any organization, whether the job involves heading a crew of lumberjacks in a forest or a police platoon in a large city. Because so many variables impinge on the day of the supervisor, he or she never really knows what emergency will develop in the next few minutes—which patrol car will break down, which citizen will complain about an officer contact, or what plan will fail.

The supervisor will find that many of the persons with whom he or she has contact want change. Some want change to improve the present situation, while others will resist such change because they are comfortable or at least familiar with the present situation, even though they recognize that it has shortcomings. Some kinds of change can be quite costly and run well past the limits of a police budget. Change also brings on the unknown, and many individuals fear that which is unknown or "different" to them. Thus, it can readily be seen that one of the important factors that helps shape the world of supervision is the anomaly of simultaneously facing conditions for change and conditions resisting change.

Some of the facets of a supervisor's environment are controllable and some are not. Supervisors may be able to control a particular subordinate, but may have very little influence over a work group when a labor relations issue is involved. Supervisors may be able to influence their bosses in decision-making problems, but may have very little influence in affecting the decisions of other supervisors at their own level. They may find themselves trapped—having a great deal of responsibility without commensurate authority. For example, supervisors may be given specific assignments to complete, but no authority to obtain the necessary additional work force needed for the job. They may have control over the work procedures of subordinates, but little control over the amount of effort that people put into carrying out procedures. Supervisors may be able to control the conditions under which their subordinates work, but may be unable to control the face-to-face contacts they have with the public. Their sphere of controllable and uncontrollable elements undergoes constant change. Thus, anyone who believes that supervisors have a great deal of control is due for surprise, disappointment, and eventual frustration.[1]

The functions of leading, commanding, and managing are basically the same, and each supervisor or member of the coordinating staff has an important relationship to these roles. However, there are some minor differences in emphasis and in primary concern.

The chief, for instance, is most concerned with mission accomplishment and the overall care of personnel and materiel resources. While wearing the chief's—or leader's—hat, he or she is likely to be more concerned with the morale and motivation of each subordinate and with the *esprit de corps* of the department.

In the manager's role, the chief concentrates more on the *efficient* use of available resources. In these times of budget and manpower austerity, the basic role of management is a constant concern for the "Four M's,"—that is, manpower, minutes, materiel, and money. This concern must apply all the way from the top leadership or management down through the chain of command to the first-line supervisor.

The Police Management Process

A police department management process is a cycle of continuing interrelated actions or functions that occur at all organizational levels. In fact, it

THE SUPERVISOR'S ROLE IN MANAGEMENT

can be looked upon as the police department's routine but dynamic system for solving problems and achieving goals. This process is comprised of twelve manager and management functions; seven of these are the functions of the manager alone—the *personal* functions to which the individual must give attention and emphasis. These seven are *"musts."* The other five are the functions of management that normally include several individuals in addition to the individual manager/supervisor. The functions of the manager and the functions of management are shown in Figures 3 and 4.

I ESTABLISHING OBJECTIVES	II MOTIVATING	III COMMUNICATING
1. Consider thoroughly what must be accomplished; distinguish between short- and long-term goals.	1. Appreciate the needs of the individual. Recognize and use individual abilities. Where feasible, have people do the work they do best, in which they are most interested, and which challenges the best within them.	1. Clarify your own ideas, desires, and purposes before communicating.
2. Examine requirements and balance them with current and attainable resources.		2. Consider timeliness and environment to facilitate communication.
3. Include subordinates in goal setting when feasible.	2. Integrate interests of the individual with those of the organization. When practicable, gain individual acceptance of organizational objectives through participation.	3. Use words that your listeners and readers understand.
4. Prescribe achievable, measurable goals, based on the above considerations.		4. See that your actions and attitudes before, during, and after your communication support your oral or written message. Encourage interaction.
	3. Recognize both good and poor performance. Be prompt, decisive, and constructive in counseling, rewarding, and disciplining.	5. See that informal communications via the "grapevine" work for the organization.
	4. Set a personal example of optimum performance, attitude, and behavior.	6. Remember, the one communicating is responsible for proper understanding by the receiver.

FIGURE 3. Actions Pertaining to the Seven Functions of the Manager

IV INNOVATING	V MAINTAINING COOPERATION	VI DEVELOPING SUBORDINATES	VII DECISION MAKING
1. Express a desire for change which will result in improvements.	1. Maintain unity of purpose throughout the organization.	1. Provide opportunities for self development.	1. Collect pertinent facts within time available. All information is rarely available at one particular time. A timely, practical decision is better than a brilliant decision made too late.
2. Foster and maintain a climate wherein change for improvement is normal.	2. Encourage freedom of communication and teamwork.	2. Clearly define subordinates' duties and standards of performance required.	2. Develop as many courses of action for consideration as possible.
3. Provide a mechanism for orderly and rapid processing of ideas for improvement.	3. Broaden individual understanding of the organization and its goals.	3. Delegate, to the extent possible, to subordinates the responsibility to make decisions. Hold subordinates responsible for these decisions and be willing to assume the risk inherent in this practice.	3. Weigh each course of action against available facts and existing conditions. Assess long-range effects of short-range decisions.
4. Provide for periodic review of policies and procedures to determine where improvement can be effected.	4. Where practicable, encourage maximum individual participation in establishing objectives and standards of performance.	4. Reward and promptly publicize outstanding performance on the part of subordinates.	4. Select preferred courses of action. Reword if necessary as a decision.
5. Have the courage to delegate and to assume the risk of failure inherent in change.	5. Balance requirements of the organization with interests and capabilities of individuals. Maintain the dignity of the individual.	5. Counsel subordinates periodically in a fair and frank manner. Stress developing subordinates' strong points and talents rather than pointing out minor weaknesses.	5. Communicate decision to those responsible for its implementation.
	6. Be prompt, decisive, and fair in action. Demonstrate attitudes and behavior which make for a cohesive atmosphere.		6. Avoid indecision. Never allow delegation to become abdication of authority.

FIGURE 3. (Cont.)

I PLANNING	II ORGANIZING	III DIRECTING	IV COORDINATING	V CONTROLLING
1. Study the situation in detail, including limitations.	1. Determine activities required to accomplish the mission.	1. Determine the extent of direction necessary by considering the type of operation, the type of organization, the experience and competence of subordinates, and the policies of top management in relation to the assigned mission.	1. Promote intelligent cooperation and mutual understanding.	1. Determine the extent, types, and methods of control necessary to keep all actions oriented toward accomplishing the mission.
2. Make reasonable assumptions.	2. Subdivide broad activities into management tasks and group-related tasks.		2. Cross-train supervisors and keep them informed as to objectives.	2. Use objectives developed in planning phase to determine realistic and appropriate standards; establish acceptable variances.
3. Review the mission or overall goal; decide on objectives.	3. Establish organizational relationships; use optimum span of control.		3. Encourage lateral and vertical communications throughout the organization.	
4. Develop an initial outline plan.	4. Select and assign appropriate personnel and other resources to accomplish the mission.	2. Consider the motivational aspects of the situation and leadership needed to secure positive action in accordance with the plan.	4. Consider time-space requirements of the plan; synchronize internal and external activities before action takes place.	3. Collect, analyze, and evaluate pertinent information.
5. Determine time and resource requirements to support this plan.				4. Compare actual results with the standards.
6. Check available resources against requirements of initial plan.	5. Allow for change in mission or resources.	3. Issue timely instructions, including when, where, and by whom each task is to be completed, and insure that these instructions are properly understood.	5. Use SOP and administrative instructions to promote coordination.	5. Take prompt corrective action to bring performance up to standard, and/or adjust norms.
7. Adjust initial plan in areas that will least affect the overall mission, balancing requirements against available resources.	6. Assign duties and responsibilities with commensurate authority.			6. Feed information back to planning function.
8. Provide for continuity and flexibility; develop alternate outline plans.	7. Emphasize essentiality, balance, cohesion, flexibility, and efficiency.			
9. Outline policies and procedures under which the plan will be implemented.				

FIGURE 4. Basic Actions Pertaining to the Five Functions of Management

It is important to understand that a police department's management process applies at every level in the department. Naturally, in the smallest subunit, the manager/supervisor will have to perform all the functions him- or herself. This is not a "Mission Impossible" task, because the scope of operation may be appropriately narrow. In contrast, at the higher levels the same director/manager/supervisor may have a whole staff section devoted entirely to supervising some functional task.

All twelve functions of the police department management process are interwoven. Although we describe them separately for training purposes, several functions will usually be performed together. For example, a director/manager/supervisor may be planning and organizing for next week's work while directing and coordinating today's work. At the same time, the director/manager/supervisor involved in the personal function of developing subordinates may be moving an officer up to a new position where the individual can produce even better work, or may be scheduling the officer for future training. Every day, director/manager/supervisors will be making decisions and, hopefully, will be motivated continuously. Motivation may require advanced techniques, or it may involve an action as simple as a pat on the back.

Principles of Organization

Organization has been defined as "the arrangement of persons with a common purpose in a manner to enable the performance by specified individuals of related tasks grouped for the purpose of assignment, and the establishment of areas of responsibility with clear-cut channels of communication and authority."[2] It is not possible to isolate principles of organization that apply to all police agencies at all times. However, no agency faces problems so unique that certain basic principles cannot be applied which have proved successful in solving somewhat similar problems in other agencies.

The International City Managers' Association's *Municipal Police Administration* offers six general principles of organization:

> 1. The work should be apportioned among the various individuals and units, according to some logical plan. (Homogeneity.)

> 2. Lines of authority and responsibility should be made as definite and direct as possible. (Delineation of responsibility.)

> 3. There is a limit to the number of subordinates who can be supervised effectively by one officer, and this limit seldom should be exceeded. (Span of control.)

4. There should be "unity of command" throughout the organization. (Subordinates under the direct control of only one supervisor.)

5. Responsibility cannot be placed without the delegation of commensurate authority, and authority should not be delegated to a person without holding him or her accountable for its use. (Delegation of responsibility.)

6. The efforts of the organizational units and of their component members must be coordinated so that all will be directed harmoniously toward the accomplishment of the police purpose. The components thus coordinated will enable the organization to function as a well-integrated unit.[3]

These principles are not the cure-all for all the organizational problems that might arise in the police agency. However, if they are applied intelligently, they can provide some answers, or at least some clues to the answers, of a large percentage of such problems. The intelligent police supervisor should realize that if sound organizational principles are to be utilized in the agency, he or she must play an extremely important role in implementing them properly.

As this is not a text on police organization and administration, there is no need to amplify these principles further. However, the modern police supervisor needs to be thoroughly familiar with them if he or she is to assume a proper role as part of the management team.

The Need for Organization

In the very small town where hazards to persons and property are minor, there may be only one police officer who attends to all law enforcement needs. Obviously, such a situation presents no problems in organization. In the police agency that employs tens, hundreds, or thousands of persons, the demands of time, local geography, different types of services called for, special problems, and a multiplicity of responsibilities make it essential that units of related work be assigned to responsible people who can be held accountable for the proper discharge of the duties of their respective commands. If the police agency is to carry out its assigned functions effectively, proper groupings of activities and designations of responsibility are essential.

The Limitations of an Organization Plan

No organization plan for a police agency should be considered a "sacred cow." A specific plan should be retained only as long as it assists the department in fulfilling the basic police purpose; it is nothing more than a convenient skeleton for the organization. An organization cannot think, solve problems, show initiative, or respond to situations; only people can. The work of the department is done by the personnel of the organization, hopefully under competent supervision, and it is only facilitated by a sound framework. (See Organization Chart in Figure 5.)

Pyramid Organization

The coordination of a department's activities is essential and can be achieved by arranging the personnel of the agency into a pyramid of authority and responsibility. Each part of the pyramid is responsible for, and has authority over, some portion of the work of the department. Every person in the department should know to whom he or she is responsible and over whom he or she has authority. In most police agencies, a chief or commissioner or sheriff is at the top of the pyramid with close assistants immediately below. As the pyramid branches out to include command-level officers, below whom are the supervisory-level officers, it finally works its way down to the patrol officer or deputy at the level of execution. Under the pyramid type of organization, cooperation between any two individuals or units in the department is possible (at least in theory) as the problem can always be referred to a higher level until an officer is reached who is superior to both individuals or units.

The pyramid type of organization has both advantages and disadvantages. As an example of a disadvantage, we cite the instance of a male department head of the traffic division who wants a beat officer to concentrate on a particular type of violation that has caused numerous accidents. First he must present his request to the supervisor who is superior to him and to the head of the patrol unit. The request must come from the common supervisor to the head of the patrol unit and then progress through channels to shift and squad commanders before it finally reaches the patrol officer for whom it is intended. An involved formal procedure of this type can waste time and promote friction between the various units. In order to cut down on the excessive red tape caused by the formal up-and-down lines of authority, too many top administrators are prone to have every division, bureau, and unit supervisor report directly to them. In a small agency this is not a serious problem, but in a large one it gives the chief administrator a span of control that is too broad to permit the job to be performed effectively.

Line and Staff Organization

The usual way to offset the disadvantages of the pyramid type of organization is for the agency administrator to establish and use auxiliary or *staff* units to provide administrative assistance. Most large police organizations use some variant of the so-called "purpose pyramid," in which a specialized unit is established to meet each of the primary functions of the

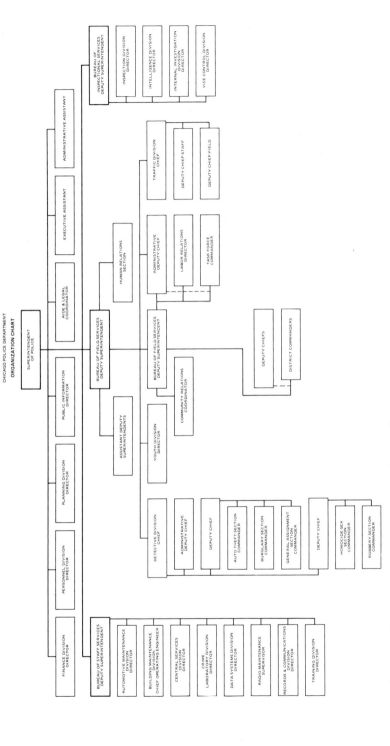

FIGURE 5. Chicago Police Department Organization Chart

department—patrol, traffic, detective, juvenile, etc. The number of the special units varies with the size of the department. These are the line units, and their work is expedited by one or more of the staff units within the agency. Staff units are usually organized by process or methods, while line units are organized by purpose. Typical staff functions in a police agency are records, communications, personnel, training, equipment, property management, etc.

A distinguishing characteristic of the staff function is that it is a service that assists the line unit in the performance of its major activities. In the small agency, line officers at some time or another perform all staff functions. However, in the large agency, the staff brings the advantages of specialization to the aid of the line organization.

Officers assigned to staff functions should never lose sight of the fact that theirs is primarily a service function, even though they might be assigned certain control duties by the central line authority. An example of the staff function is the report review officer who inspects reports submitted by line officers for completeness, though the officer's position is that of a records officer. Properly, an incomplete report would not be routed directly back to the writer, but rather would be routed through the line supervisor.

Do's and Don't's for Supervisors Assigned to Staff Duties:

> 1. Do make full use of the line officer's knowledge of operations, and remember that consultation will promote cooperation.
>
> 2. Do not do anything to undermine the line position; your job is to assist the line unit so that it will be able to do the best possible job.
>
> 3. Don't give staff advice in such a manner that it will be interpreted as a line order.
>
> 4. Don't give staff advice without checking it first for accuracy. It is a bad mistake to count on the line to catch staff errors.

Do's and Don't's for Supervisors Assigned to Line Duties:

> 1. Do remember that you and the staff officers are working toward the same goal, which is the accomplishment of the police mission, and that you can help each other to attain that goal.
>
> 2. Do take full advantage of available staff services. It is a terrific waste of time to work on details when there is a specialist available to handle such matters.
>
> 3. Don't try to pass the buck to the staff unit. Remember that you are responsible for actual operations.
>
> 4. Don't confuse advice with an order. They are two *different* things.

The need for the police supervisor to understand organization principles is basic. The unit supervised is an integral part of the larger organization, and the supervisor is primarily responsible for organizational structure within the unit. Even though the unit structure may have been established by other persons, the supervisor is in a position to recommend changes and improvements if a strong case can be built for them.

Most supervisory personnel hope to advance eventually to command positions. Today, a knowledge of the fundamentals of organization is an increasingly important factor in the selection of police executives. Because of the great complexity of the problems faced by modern police agencies, the supervisor who has the knowledge and the ability to solve these problems is the person most likely to be selected to fill a higher rank. For this reason, the good supervisor will prepare for promotion by studying the principles of organization and management and analyzing them in terms of the individual's own department.

Records—A Basic Supervisory Tool

The supervisor who tries to remember too many details handicaps performance unnecessarily and will find that he or she lacks factual information when it is needed. A good records system should contain all of the facts necessary for the successful operation of the police agency. The term "records," as used here, refers to all auxiliary as well as operational records kept by the agency.

The preparation and use of adequate records take the guesswork out of management. For example, a chief of police who requests funds for additional manpower from the city council must have sufficient records to prove department needs. Inadequate records can be quite costly, but remember that too many records may be equally costly.

Some of the records that are most useful to supervisors are those compiled in other units of the department. Records maintained by the department's personnel division are utilized by the supervisor for work appraisal, promotion, complaints, and discipline.

Consulting records periodically to see how one's unit compares with other units in the department is very helpful. This procedure enables the supervisor to locate problem areas and may give clues as to what measures to take in eliminating them. The records of the traffic division detailing the location and frequency of accidents can be utilized by patrol-division supervisors to pinpoint problem areas for concentrated patrol effort.

The good supervisor controls records, but does not allow records to be a controlling factor. The supervisor cannot supervise a unit effectively if most of his or her time is spent at a desk handling record-keeping tasks. Even though there are some records which must be kept personally, particularly those used to rate subordinates, the supervisor cannot be buried under a mound of paper.

Records often serve as an instrument of supervisory control. A check of case reports and statistics on job distribution allows the supervisor to allocate work in an equitable manner so that no person or unit is overburdened while another is idle.

There is no question that subordinates will resent having to prepare the records used in setting up controls if they don't understand the need for them. Most people dislike being checked on and will rebel at having to prepare a report that may show them failing. Also, most police officers are fairly active individuals and do not particularly enjoy writing reports of any kind. Too many officers have a tendency to handle certain matters "off the cuff" instead of including them in a report. A good supervisor will impress upon subordinates the importance of good reporting procedures.

One recent incident illustrates the importance of records. Late one night, a sheriff's unit was dispatched to the scene of a disturbance, but was unable to discover anything. No record was made of the incident, nor was the name of the complainant recorded. Even the phone memorandum was thrown away. The importance of the incident was established rather vividly the following day when passers-by found a partially dismembered body in a roadside ditch at the same location.

Generally, the attitude of the supervisor concerning report-writing and record-keeping will be reflected in the attitudes of subordinates. If the supervisor displays an attitude of laxity in the enforcement of departmental requirements, subordinates will feel little compulsion to comply with those requirements. Such laxity creates a problem for the new supervisor who takes over the leadership of a group, as any sudden attempt to tighten up will be resisted by lower-ranking personnel.

Delegation of Responsibility

Because no top police executive can begin to carry out all of an agency's managerial functions, the practice of all good executives should be followed and some duties and responsibilities should be delegated to other people. Proper delegation of duties and responsibilities requires that the executive give the individuals the necessary authority and backing to do the job, supply these people with the required personnel and materials, and see to it that the delegated functions are carried out as directed. The individuals to whom these responsibilities are delegated are quite likely to be overburdened themselves, and they will soon delegate a portion of their assigned tasks to others and see that they are done in accordance with instructions. This delegation of work continues down the chain of command to the level of execution—the officer or deputy who does the basic work of the department.

Difference Between Management and Supervision

Because the first-line supervisor gets results by working with and through people, he or she is a member of management. This job differs principally from that of the top police executive in two ways.

First, the scope of the supervisor's duties and responsibilities is much more restricted. Much less time is devoted to planning, coordinating, and policy formation, and much more time is devoted to activities such as training, developing, directing, and energizing the work of subordinates.

Second, the supervisor directs the work of subordinates who actually do the job of carrying out the police purpose, whereas top management directs the activities of other executives and supervisors.

Written Statements of Duties and Responsibilities

Each supervisor must have a clear understanding of duties and responsibilities. Too frequently, it is assumed that the supervisors know what their duties are and have a full understanding of their responsibilities. How can this possibly be the case when departments have failed to provide a written manual of policies and procedures? Each department should have such a manual stating the duties and responsibilities of every person in the organization, from the lowest employee to the top administrators. Furthermore, the employee, supervisor, or administrator should participate in the drawing-up of the statement of their own duties and responsibilities and have a copy of it for personal use.

The statement of duties and responsibilities should indicate clearly the individual supervisor's relationship to other supervisors in the same department, and to the staff and operational divisions. These include the records and identification, personnel, training, detective, and patrol divisions, among others.

A job analysis is necessary to clarify further each supervisor's duties and responsibilities. Each supervisor should categorize his or her duties and responsibilities into "jobs performed" and classify these jobs by the type or types of responsibility they entail. Such a job analysis is often referred to as a "laundry list."

Organization, Division, and Function Charts

Each supervisor should also have an organization chart showing the individual's position in the management structure, and should have a definite picture of the line organization and of the channels through which management actions flow. This chart will minimize the possibility of conflicting responsibility and overlapping authority.

In addition, each supervisor should prepare the following guides: (1) a chart showing the structure of his or her specific division of the organization, and (2) a chart showing the functional relationships of this division to the other divisions of the department.

Definition of a Supervisor's Authority

Delegation of Authority

Delegation of authority to the first-line administrator should be specific and should be included as part of the written statement of duties and responsibilities. The authority should be sufficient to permit the supervisor to carry out administrative responsibility. The authority delegated is of three fundamental types: full authority, limited authority, and approval before taking action. The statement should specify which type is involved in each instance.

Full Authority   This authority applies to functions or situations in which the supervisor (within the framework of the department's policy), without consulting or getting clearance from a superior, controls such activities as: assigning work to employees, granting time off, requisitioning routine supplies, taking disciplinary action within prescribed limits for infraction of department regulations, and authorizing overtime work.

Limited Authority   This authority is the same as full authority, but requires a prompt report of action taken. Limited authority is granted for functions such as: suspension of employees for inefficiency, insubordination, or suspicion of illegal acts; emergency situations that require immediate action to avert hazards to the public, personnel, or property; and authorization of changes in work methods or details of programs necessary because of unforeseen conditions where failure to act promptly will cause unreasonable delay.

Approval before Taking Action   This authority applies to those functions or situations in which the supervisor is expected to make recommendations that *must be approved* before taking action. The supervisor requests approval for actions such as the discharge of an employee, deviation from established policies or procedures, promotions or demotions, and the modification of established operating practices.

Areas to Be Included in Each Supervisor's Authority

Specific areas should be included in each police supervisor's authority.

Transfers   Whenever feasible, the supervisor should have the right to approve or disapprove the transfer of an officer from one position to another.

Whenever practical, the supervisor should have the opportunity to interview a new officer before that person is assigned to the supervisor's unit.

Planning and Organizing   Each supervisor should have the authority to plan and organize his or her work *within the limits of the department's policies and established operating procedures,* and should be held responsible for the results obtained.

Part of the Administrative Team   The supervisor's authority as a responsible part of the administrative team must be firmly established in the minds of subordinates. The supervisor's status and prestige must be apparent from the attitude and actions of superiors toward the supervisor. If superiors do not respect the supervisor's position and judgment, the subordinates will not. If the supervisor is to retain prestige and authority in the eyes of those supervised, this person must be their first contact for handling grievances, complaints, misunderstandings, individual counseling, and requests for information. It is essential that the established grievance procedure emphasize this fact. To bypass the immediate supervisor without this person's knowledge in the handling of a grievance, misunderstanding, or any other personnel problem lessens the authority of the supervisor and weakens his or her ability to carry out the function as the employment-relations authority. The handling and preventing of complaints and grievances is treated in detail in Chapter 13.

Duties and Responsibilities of the Police Supervisor

In reviewing and analyzing the characteristics of a supervisor, one finds that this is an individual who directs and controls the work of others. Therefore, it should be recognized that this position has a great deal of responsibility. Only through the continuous study of supervisory responsibilities will the high quality of supervision that is essential in a modern police department be developed and maintained. To give a better understanding of this important part of the supervisory position, it is essential to list some of the common duties and responsibilities. Of course, the supervisor must have technical knowledge, but primarily must be a good manager, capable of running a particular unit or division. Regardless of rank—sergeant, lieutenant, or captain—a supervisor has three main responsibilities: administration, supervision, and training.

Administration includes: Those types of activities which are concerned particularly with planning, organizing, staffing, directing, coordinating, recording, budgeting, and public relations.

Supervision includes: Techniques employed to get the job done; the checking of results and the determination of the causes of success, failure, or mediocrity; and the control, development, and maintenance of harmonious relationships among all employees.

Training includes: Instruction in the development of good habits and attitudes; practical instruction: how, what, when, where, and why tasks are to be done; and the development of potential in subordinates.

Administration

Planning   Although the primary responsibility for planning rests with the department head, it must be carried out at every level within the department chain of command. Supervisors must plan for the implementation of departmental policies and tactical operations, manpower requirements, and beat configurations, and must sell the department's plans to subordinate personnel. Supervisors also must plan training for roll-call, in-service, and special training sessions.

Organizing   The supervisor must organize subordinates, equipment, and tasks in such a manner that the organization's goals will be achieved with maximum efficiency and minimum expenditure of personnel, time, and money.

Staffing   It is the supervisor's duty to make work assignments. These assignments are determined by the available manpower, special problems, the time of police events, the size and type of area to be covered, and the type of work to be accomplished. The supervisor must deploy subordinates in a logical and efficient manner.

Directing   The essence of leadership is the effective use of the command, the request, the suggestion, and the volunteer. Each of these can be used to advantage, depending upon the circumstances and the persons involved. Directions should include who, what, when, where, why, and how. In emergency cases the explanation of *why* must wait until the mission is concluded. Directions must be clear, concise, and understood by those who are expected to respond. Subordinates should be allowed to act upon their own initiative, but the limits of their authority must be defined.

Coordinating   Supervisory reports are required at all levels of supervision in order to insure coordination. As Clifford L. Scott has said: "The flow of reports from the field may be compared to the sensory nerves which supply sensation, information, warning, and all other forms of intelligence to the brain.

"The efficient production and processing of reports is a leadership problem of considerable concern. Every supervisor bears the responsibility for prompt, accurate, and complete preparation and processing of reports."[4]

Some of the types of reports are: unusual events, problems, general information, performance evaluation, statistics, and documentation of critical events. Reporting is necessary in order to compile information that may affect staffing, deployment of personnel, patrol area configurations, use of, or need for, equipment, production standards, and comparison of personnel performances.

Budgeting First-line supervisors may not be directly involved in the preparation of the department budget; this responsibility generally falls within the purview of a command officer. But the supervisor plays a definite role in determining the department's needs and must prevent waste of supplies and misuse of time, personnel, and equipment.

Public Relations Public acceptance is the key to the survival of a police organization. The only thing the police have to sell is service. If public support is lacking, the agency's effectiveness will be minimal. Good public relations can be established by the interest shown by employees in the individual citizen's problems, the quality of information given to citizens, and the employee's mannerisms, speech, and appearance. Press relations are of prime concern and, if mishandled, the resultant damage can be fatal to the organization.

Supervision

The supervisor has been described as "any person who is responsible for the conduct of others in the achievement of a particular task; for the maintenance of quality standards; for the protection and care of materials; and for services to be rendered to those under his control."[5]

In a narrow context, supervision is the acceptance of responsibility for the accomplishment of an assignment in an acceptable manner by subordinates through the use of a variety of techniques. Pure supervision involves directing, inspection, follow-up, and control. The broader responsibilities of supervision involve handling matters that pertain to the welfare, the interest, and the interrelationships of the subordinates. Some supervisory tasks entail: assigning personnel to jobs and specific tasks, watching the results, reporting feedback to superiors, taking disciplinary action, and evaluating subordinates and correcting their errors.

Training

All supervisors have a training function to perform. The first- *and* second-line supervisors are interested primarily in everyday training, such as briefing, in-service training, and performance rating. They teach the how, what, when, where, and why of the job. They give information on staff policies and procedures. It is their responsibility to develop in their subordinates good habits and attitudes. They must prepare lesson plans, conduct training sessions (which should include disciplinary training), and institute other types of training programs that will produce high-quality employees.

Principles of Responsibility

Several rules that govern responsibility must be made clear:

1. Final responsibility must not be divided.

2. Responsibility must be accepted fully or not at all.

3. Responsibility may only be delegated, not relinquished.

4. Responsibility should always be accompanied with commensurate authority.

5. The person who assumes a responsibility should know what is expected and realize that he or she alone will be held responsible for the results.

6. The expected results should determine the conditions of delegated responsibility.

7. The method of carrying out tasks should be left to the person who assumes the responsibility.

8. No authority should be assumed in areas in which a supervisor is not responsible. Each person should stick to his or her own job.

The Supervisor and Waste Control

One of the never-ending battles of police management is that of staying within an operating budget. Thus, control of waste is one of the most important management functions performed by supervisors. Waste may occur in the following areas: manpower, equipment, supplies, materials, heat, light, and power. It may be classified as avoidable, semi-avoidable, and unavoidable.

While management is responsible for many aspects of waste control, management must rely on its representatives for the actual control at the operational level. Thus, the supervisor becomes the key person in waste reduction. Some waste results from management decisions, and some is caused by subordinates.

Waste caused by management results from poor planning, or lack of planning, from faulty organization, inadequate facilities, poor layout, inefficient methods, lack of controls, inadequate instruction, improper use of employees' skills, etc.

Waste is caused by subordinates when they fail to acquire the necessary skills to do a job, lack the willingness to cooperate, or have the wrong attitude toward their work, their supervisor, or the department. Some deliberately "dog it" on the job, fail to learn and to adhere to the standards set for them, have accidents they could have avoided by following proper procedures, or fail to pay the necessary attention to details.

The first step in waste control is to get all the facts, the second is to determine the causes and fix the responsibility, and the third is to take whatever action is necessary to control or eliminate the waste.

In combating waste, the supervisor must set up adequate controls. An example of such a control is the plan for and development of methods to care for department equipment and materials. The elimination of waste requires that equipment be properly used and periodically inspected. The supervisor should make use of records and statistics to control manpower waste. Personnel should be placed where they can make the best contribution to the police effort; the necessary steps must be taken to insure that the men and women put forth their best efforts. Waste control can be achieved through planning, close supervision, training, and cooperation. The following is a detailed discussion of each type of waste control.

Waste of Personnel.

Generally, every supervisor who makes an analysis of his or her unit finds areas where personnel are not being utilized efficiently. In some instances, the personnel are not using the best work methods, while in other cases, the supervisor finds that the personnel are doing jobs for which they are not qualified. A skilled individual might be doing work that could be performed by an unskilled person. This kind of waste occurs quite often in police work, particularly in such areas as records and parking meter enforcement. Personnel waste also occurs when too many officers are assigned to an area of low crime incidence or where called-for services are at a low level. The selection and placement of personnel should be continually controlled; individuals should be trained or transferred when the placements are unsatisfactory.

Waste from Accidents

Waste quite often results from accidents which decrease the output of the unit and force the remaining personnel to assume a greater load. Financial losses are suffered by the department through compensation costs and drains on the pension funds.

Waste of Equipment, Supplies, and Materials

Tremendous dollar losses result from the misuse of equipment and the waste of supplies and materials. Simple matters such as improper operation of police vehicles can cause huge increases in maintenance costs. An officer who forgets to check the oil level in a patrol car before starting a tour of duty can cause damage requiring a complete engine overhaul. Multiply this expense by the number of officers operating police vehicles and the potential cost is staggering. Frequent "jackrabbit starts" can damage automatic transmissions, resulting in more "down time" for the vehicles and higher costs to the department. Misuse of typewriters and other office equipment also results in considerable waste.

Countless examples could be cited to show where efficient supervisors could cut down on waste by controlling the misuse of equipment, supplies, and materials. Even though the cost of a single item might seem minor, if it is multiplied by the number of persons in the unit its importance becomes obvious. It is clearly the responsibility of the supervisor to keep such waste to a minimum.

Waste of Heat, Light, and Power

This type of waste in most police agencies falls properly within the realm of building maintenance; however, even in this area, the line supervisor should make recommendations that human controls be replaced with automatic controls whenever possible.

Summary

The supervisor should study carefully the duties and responsibilities of the job and know the framework in which he or she is expected to operate and the relationship of supervisors of other units to this framework. A sure knowledge of the boundaries of responsibility is likely to reveal some duties of the unit which are being neglected and others that are being overemphasized. Knowledge of the limits of responsibility will keep a supervisor from assuming responsibilities that have not been assigned. In terms of personal responsibility, the supervisor will be able to see more clearly that subordinates and equipment are being used to their best advantage. Additionally, the supervisor will be able to coordinate and plan activities more effectively, and will be able to distinguish between those duties which can be delegated and those which cannot.

Management functions may be divided into two broad categories: those functions which require contact with subordinates, and those of an impersonal nature. The following list, which is divided according to this classification, summarizes the duties and responsibilities that have been discussed in this chapter. Although it does not include every function that a supervisor may be expected to perform, it does include the most important ones. Management functions involving contact with subordinates are:

1. Orienting new employees
2. Interpreting departmental policies and regulations for subordinates
3. Training personnel to perform their jobs safely and efficiently
4. Giving orders and assigning duties and responsibilities
5. Maintaining discipline
6. Handling complaints and grievances
7. Building morale
8. Keeping subordinates informed of proposed changes that will affect them, and of their status in the organization
9. Looking after the comfort of subordinates
10. Developing subordinates' rotating assignments, training, etc.
11. Securing cooperation from subordinate personnel
12. Assisting subordinates who have personal problems

Management functions of an impersonal nature are:

1. Planning personnel work
2. Coordinating the activities of subordinates with the work that must be done
3. Inspecting work, materials, equipment, etc.
4. Seeing that equipment is maintained
5. Seeing that working conditions are kept up to standard
6. Maintaining a high quality standard for work
7. Safeguarding the health and safety of subordinates
8. Eliminating waste
9. Keeping records on work output, case loads, etc.
10. Preparing reports
11. Cooperating with other supervisors and command personnel
12. Assuming responsibility for errors and mistakes of subordinates
13. Keeping superiors informed about the work, the attitudes, and the desires of subordinates

DISCUSSION QUESTIONS

1. Why is it so important that a police agency have a written statement of duties and responsibilities?
2. What is meant by full authority?
3. What is meant by limited authority?
4. What is meant by approval before taking action?
5. What is meant by each of the eight types of supervisory responsibility?
6. Define the term "organization." What is its importance?
7. What are the six principles of organization? What do they mean? Why are they important?
8. What are the weaknesses of the simple pyramid type of organization? What are the methods used to overcome these weaknesses?
9. Differentiate between line and staff functions. How do they supplement each other?
10. How are records used as a tool of supervision? Why are they important?
11. What are the management responsibilities of supervisors?
12. In what ways can the supervisor cut costs through the elimination of waste?

NOTES

1. Joseph L. Massie and John Douglas, *Managing: A Contemporary Introduction* (Englewood Cliffs, N.J.: Prentice-Hall, Inc., 1973), pp. 7-8.
2. International City Managers' Association, *Municipal Police Administration*, 5th ed. (Chicago), p. 44.
3. Ibid., p. 45.
4. Clifford L. Scott, *Leadership for the Police Supervisor* (Springfield, Ill.: Charles C. Thomas, 1960), pp. 95-96.
5. William R. Spriegel, Edward Schulz, and William B. Spriegel, *Elements of Supervision* (New York: John Wiley and Sons, Inc., 1967).

4 The Selection of Supervisors

Selection of police supervisors cannot be based on a hit-or-miss method, nor can it be left to chance. There is no system that will unfailingly select good police supervisors; however, the procedures utilized are constantly being refined and revised in order to reduce the element of chance as much as possible. Few supervisors are given much real help in the development of their ability to plan and organize work or in the construction of good techniques of handling their subordinates. In too many police agencies this development is left to chance. It is seldom possible to select individuals who already possess these skills; thus, it is best to select people with high potential and then to help them to utilize that potential to its fullest extent.

The aim of any good supervisory development program is to select the persons best qualified to meet the requirements of the supervisory positions to be filled. The successful achievement of this goal requires that police management specify the requirements of the positions, determine the qualifications of the potential candidates, and select, on an objective basis, the best qualified of the candidates. It is important to know something of good selection techniques; but in order to understand such techniques fully, it is necessary to be able to identify the qualities needed in a person who is to lead others.

Qualifications of Supervisors

Police officers often complain that the promotion system in their department is unfair—that they spent considerable time preparing for the examination, but found only a small number of questions pertaining to the technical knowledge required for a supervisory job. They also complain that most of the questions concerned subjects with which they were unfamiliar. Some such questions might be, "An important departmental control is _____?" or "If you were confronted with a disciplinary problem, e.g., _____, which of the following methods would you use?"

Of course, such officers probably prepared for promotion the same way their predecessors did—they relied basically upon the same ideas that succeeded for the old-time supervisor, "the old boss."

The "old boss" type of supervisor, who was predominantly male, usually forced others to perform tasks by threat or even by the use of his fists. He had to be tough enough to get away with it. He displayed job knowledge and skills by getting his work out rather than by his ability to handle people. Any means employed were satisfactory, as long as they produced the desired results. The old boss did not encourage suggestions or ideas from his subordinates, and when they were offered, he often stole them to use for his personal advantage. His subordinates produced results because of fear, not because of a desire to get the job done as part of a group effort.

Quite different qualities are required of the modern supervisor. The promotion examinations of today are directed toward finding individuals who have different qualities and experiences than those of the "old boss." In recent years, police management has been forced to recognize that the primary qualifications for supervisory personnel have undergone a considerable change since the days of the "harness bull" and "old boss." Some of the qualities needed by a good supervisor are: (1) ability to handle men and

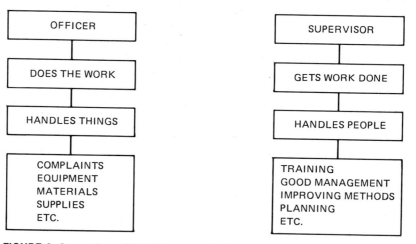

FIGURE 6. Comparison of the Duties of a Patrol Officer and a Supervisor

women, (2) ability to plan, (3) ability to organize, and (4) technical knowledge. A supervisor must be able to understand and apply departmental policies and procedures. The supervisor must be loyal to the department and to subordinates. It is essential that supervisors be able to handle people by building them up instead of tearing them down, and that they be able to take care of many complex responsibilities—while getting out the work. A supervisor must have the technical knowledge needed to understand and direct complicated operations and be able to encourage others to produce ideas and put those ideas to work.

Certain qualities necessary for a competent supervisor have been established by conferences with highly rated supervisors.[1] These conferences have brought out the following characteristics: (1) on-the-job appearance—a confident and inspiring attitude; (2) firmness—persistence and decisiveness; (3) technical knowledge—thorough job knowledge and superior intelligence; (4) teaching ability and enthusiasm; (5) reliability, integrity, dependability, and loyalty; (6) sympathy and patience; (7) desire to learn; (8) cooperativeness and ability to inspire the respect of others; (9) tact, emotional stability, moral courage, and a sense of humor; and (10) fairness, friendliness, and understanding.

Modern Procedures for Selection of Supervisors

Supervisory personnel must be chosen in a systematic manner. Many management studies have established that the casual methods used in the past are in the long run costly and of doubtful validity. They are costly because of the high number of mistakes that result. A single lawsuit for false arrest resulting ultimately from a poor decision by a supervisor can cost enough to underwrite for many years the cost of a good selection system.

It has been demonstrated repeatedly that when the sole method of selecting supervisory personnel is nomination by superior officers, well-qualified people will be passed over if they have not drawn attention to themselves. The traits that often draw a supervisor's attention to a subordinate—extroversion and aggressiveness, for example—do not necessarily have a high correlation with other traits that are desirable in the police supervisor.

A systematic method of selection will bring into the pool of talent competent persons who might have been ignored by other methods of selection. Selection of supervisors based on merit is the only system which will not destroy morale, and this is most important in any police agency. Hit-or-miss methods will result in complaints of favoritism from individuals who have not been chosen for promotion. A strong merit system of promotion utilizing good tests as part of the basis for selection will tend to eliminate "halo" factors. People who are competent, rather than flashy, will emerge from the crowd.

A clinical approach based on thorough consideration of all factors is essential to a good selection process. This approach might utilize techniques similar to the highly specialized ones used in the academic and professional disciplines.

Data should be collected about candidates for supervisory positions. Collecting data includes obtaining information about the candidate in areas such as the physiological, psychological, and sociological. In other words, the individual's health, home, family, intelligence, work history, and temperament should be known. Therefore, there is a need for some system for collecting, filing, and analyzing such data in an orderly and systematic fashion. Depending upon the size of the jurisdiction, this job may be handled by the city or the county personnel department or by a personnel unit within the police agency. In either case, the unit must have imaginative and creative leadership, and must not be relegated to conducting a routine clerical operation.

The supervisory selection program must be comprehensive in its methods. There is no known "single-shot" device or approach to selection that will take care of all the problems involved in picking good police supervisors; however, a comprehensive selection program will include some or all of the procedures and considerations discussed on the following pages.

Written Examinations

No professional psychologist would categorically favor any single device for measuring supervisory or administrative ability. A good selection system without tests would be hard to devise, but it must be remembered that, at the present time, no single test has been devised that is sufficiently precise to serve as the entire basis for the selection of police supervisory personnel. Tests should be validated within the local department, as they have a way of working in one environment and not in another. This does not mean that tests are of no value; it merely means that each test must be tailored to do a particular job for the agency in which it is to be used.

A test is said to be *reliable* if the same results are obtained in repeated administrations of the test to the same group of testees, and if the same thing is measured when the test is given repeatedly to similar groups of testees.

A test is said to be *valid* when it actually measures the traits, aptitudes, or skills that it purports to measure. A validation study is an effort to determine if a test actually works as expected. To validate a test, it is administered to a criterion group which is subdivided into a control group and an experimental group. If we wish to find out if a given test is valid as a differentiator of high- and low-performance police supervisors, we administer it to groups in each category. If supervisors in the high-performance group make significantly higher scores than those in the other group, we can assume that the test is valid within certain statistical limits.

Tests should also be subjected to an *item analysis,* which is essentially a validation of each item in the test by a procedure similar to that used for the entire test. The purpose of item analysis is to determine which of the individual items or questions are carrying the testing load and which are merely serving as "padding." If a particular item or question does not discriminate between the experimental and control groups, it should be eliminated from the test or modified until it is discriminatory.

Factor analysis is a technique used to compare test items and groups of items for economy—that is, to determine if the items are measuring the same things. Items or groups of items that are measuring the same factor can be identified and consolidated or eliminated. This is called *purifying* the test.

Certain types of personality tests are considered highly controversial in the police service; these tests discriminate by psychological tendency. In effect, the testees are placed in a psychological category—schizoid, paranoid, etc.—which represents their basic personality tendency. The police supervisor should realize that these tests are controversial, and, should they be utilized in his or her department, the tests should not rely totally on the results obtained. Evaluating police officers for promotion is a complex problem that must be approached with a battery of tools rather than a "single-shot gimmick."

The Oral Interview—"Qualifications Appraisal Boards"

Most police agencies today are using the oral interview, or some variation of it, as a part of the supervisory selection process. It is felt that a board that is properly set up can size up the individual candidates for possible promotion to supervisory positions. In some jurisdictions the oral board has been used as an administrative device to control the selection procedure; however, this is not the purpose of an oral board, and it has been repeatedly demonstrated that a properly constituted board can do a fair and effective job of selecting the best qualified candidates for a position. The board is in a position to evaluate certain qualities, such as appearance, voice, tact, ability to stand up under stress, etc. These qualities cannot be evaluated entirely by other means.

There is a good deal of debate about the composition of the board. Some administrators feel that the board should be made up entirely of administrative and supervisory personnel from within the department. Others feel that such a board would not be fair to some of the candidates appearing before it and could too easily be completely dominated by the head of the agency. At the opposite end of the scale, some of those in the personnel field believe that the board should be composed entirely of people from outside the agency in order to ensure its complete fairness. However, the argument against this type of board is that it has no knowledge of local problems or supervisory personnel requirements. Probably the best solution to the problem is a balanced board composed of both local people and outside members. The California State Personnel Board has had a good deal of success with the three-member board, composed of one representative from the agency involved, one from the Personnel Board itself, and one member of the public who has some familiarity with the requirements of the position to be filled. Generally, boards vary in size from three to seven members; but regardless of size, all members should have a good deal of skill in interviewing so as to be able to detect relevant characteristics—good and bad—of each candidate.

Local procedures usually govern what *areas of inquiry* the board may explore and the extent of such an inquiry. In many jurisdictions, an oral board may not ask any questions about job knowledge, as it is assumed that this area has been covered adequately in the written examination. Most jurisdictions require that boards avoid questions pertaining to religion or politics, for obvious reasons. However, an oral board is usually given sufficient latitude to set up hypothetical situations so that they can gain some insight into a candidate's problem-solving ability and reaction to artificially induced stress.

Tests of Social Skills

Social skills are among a police supervisor's most important assets. How does the candidate get along with other people—superiors, subordinates, and peers? Does the candidate inspire confidence in others, and do other people like to be with this person? Does the candidate emerge as the natural leader of a group, and is good judgment shown in dealings with other people? Is this person able to maintain social equilibrium and emotional stability under conditions of stress and provocation? These are some of the questions that must be answered.

In recent years, a new method of measuring social skills has been developed which has been given extensive use in civil service testing. This is the *group oral interview,* which has gained popularity in recent years and is now being used fairly extensively in police supervisory selection, sometimes alone and sometimes in combination with the regular or conventional oral board.

The group oral interview, or "leaderless discussion," as it is known to the social scientists, is a device for evaluating the potential of persons who aspire to become leaders. Typically, six or eight candidates are placed around a conference table where they can be observed by the raters. They are given a question to discuss or a problem to solve and then left on their own. No member of the group is designated as the leader, because one of the objectives is to see which of the candidates will demonstrate qualities of leadership in the group situation. Sometimes a time limit is assigned to see how well the group can organize itself. Some agencies follow the group interview with individual interviews in which the questions asked have been suggested by performance in the group setting.

Service Ratings

The ratings of a candidate's superiors should always be appraised carefully when considering the individual for advancement to a supervisory position. However, the reliability and validity of performance ratings is dependent upon the manner in which the supervisors of the program approach the job of rating. If supervisors approach the rating program objectively, using a painstaking, clinical approach, allowing adequate time for it, and incorporating it into their training programs, then it should be a part of the promotional process. This method requires that the supervisors put their whole heart into the program and have the courage of their convictions. What sometimes happens is that the rating program is given only lip service; everyone is rated either high or low, or the rating program simply degenerates into a popularity contest. If this occurs, the ratings should obviously not be used when selecting supervisors. Furthermore, the rating being used for candidate evaluation should not be just the latest rating received by the candidate, but rather should be a composite of all the ratings received over a specified period of time prior to the examinations. A two-year period usually is considered adequate if ratings are given regularly—at least twice each year.

Seniority

Seniority is utilized by some agencies as one of the criteria for promotion. The main argument in favor of this practice is that older personnel are entitled to some consideration because of their years of service to the department. This argument becomes invalid when one accepts the premise that the best available person should be selected to fill any given position. Length of service beyond the minimum necessary to learn the appropriate job skills has never been shown to be a valid criterion of supervisory ability. The seniority system has the further disadvantage of destroying the incentive of the younger personnel in the department. No matter how competent the person might be, a promotion must wait until all of the department's "graybeards" have been promoted. This system usually results in unrest, low morale, and high turnover rates among the younger people in the agency. Seniority, therefore, should not be considered in the selection process.

Peer Ratings

"Peer ratings," or the rating of an individual by co-workers, have been little used by police agencies. But there is no doubt that a person's fellow workers are in the best position to evaluate job performance. One of the highest praises that can be given to any police officer is when a fellow officer says, "He (or she) is a good cop." We all have had the experience of working with fellow officers whom we knew we could count on when the going got rough, and most individuals who have been in the profession for any length of time have also had the experience of working with people whom they knew they could not count on when the chips were down. The police service should draw upon the experiences of business and industry in this area and at least explore the possibilities of using peer ratings when considering individuals for promotion. A five-point rating scale ranging from poor to outstanding could be used to evaluate various aspects of an individual's potential to become a good supervisor. Areas that might be rated by peers include: (1) job knowledge, (2) work performance, (3) leadership ability, (4) contribution to morale within the group, (5) self-control, and (6) ability to instruct or give directions. Ratings in these specific areas should be followed by an overall appraisal.

Assessment Centers

One of the newest systems used by the police service to select managers and supervisors is the *assessment center*.[2] An assessment center can be visualized as a system which accurately and fairly judges people: in this case, police personnel who are being evaluated for promotion. The methodology spans the entire gamut of entry-level selection, performance appraisal, skill development, and promotion at all levels. Defined somewhat more precisely, an assessment center provides a comprehensive and in-depth situation-based method for improving a manager's accuracy in evaluating the capabilities of existing or potential staff. In this context, "situation-based" means a reality-oriented demonstration of job-related skills which must be validated, and an observable process for generating behavior and exposing skills to trained observers or assessors. Historically, assessment centers can be traced to the early 1930s, but the first known use in the police service was in 1972 when the city of Riverside, California, used the process for selecting lieutenants. Since that time, the International Association of Chiefs of Police, Justice Research Associates in Costa Mesa, California, and other groups have been providing assessment center services to police agencies.

Typically, an assessment center's program will involve from one to three days of intensive work with the candidates and a ratio of one assessor to each two to five candidates. The candidates will be involved in role-playing exercises, standardized tests, management and supervisory exercises, leaderless group discussions, analytical problem solving, fact-finding, and other exercises that are related to job requirements. The assessors then collectively analyze their data and rank the candidates. Discussions by the authors with personnel who have been through an assessment center have revealed a high degree of satisfaction with the results, even from those who were not highly ranked by the assessors. While this technique is still relatively new for the police promotion process, it appears to have great potential as an improvement over traditional civil service methods of selecting supervisors.

Developing Potential Supervisors

Since World War II, industry has recognized more and more the importance of the supervisor's role in the organization, and, as a direct result, much more attention has been devoted to the development of better supervisory leadership. Many companies are also recognizing that special care must be taken to select those individuals who have the greatest potential and who can most benefit from special training in supervisory attitudes and skills. Most large companies have established some form of supervisory development program similar to those designed for middle and top management. The smaller companies tend to encourage supervisors and potential supervisors to obtain outside training that will enable them to become leaders.[3]

Unfortunately, the police service has lagged behind business and industry in establishing planned programs of supervisory and executive development. However, this type of program is beginning to make an appearance in police service and will undoubtedly appear more often in coming years. A survey by Allen Bristow revealed that, of the representatives of 257 California law enforcement agencies who replied to his questionnaire, 30 required that candidates for promotion to the rank of sergeant have completed at least some college-level courses. The requirements varied from completion of nine college units to the attainment of a bachelor's degree. Twenty-one departments gave a salary bonus for completing college or for semester units taken; sixty-three agencies provided various types of compensation for college attendance. Bristow concludes:

> It would seem that California law enforcement agencies are establishing recognition for professional college education. The information presented in this article is based on a 75% return of the questionnaires, and it is assumed that the requirements of a number of agencies are not included.
>
> It should be noted that the largest agencies in the state do not recognize college education in a formal manner. This may be misleading, however, because the personal observation of the author would indicate college education on these departments is so universal that recognition would have little effect on professional attainment. It has been said of these agencies, "you can't make sergeant without a degree, not because of regulations, but because of competition . . . everyone else has one!"[4]

A number of years ago, recognizing the value of criminal justice education at the college level, the Sacramento County (California) Sheriff's Department changed the time of service required for eligibility to take the sergeant's examination for those officers holding a bachelor's degree. The person with a degree is eligible for promotion after three years of service as a deputy, while the individual without a degree must serve four years before a promotional examination may be taken. In those states that have established a commission on peace officer standards and training, a number of the commissions have established some form of basic, intermediate, and advanced peace officer certificates based on varying combinations of experience, education, and training. (See appendix for California certificate requirements.) A number of the departments in these states are requiring that promotion candidates hold either the intermediate or the advanced certificate before they become eligible for promotion. It is the prediction of the authors that in time this will become a fairly standard statewide requirement in those states that have commissions on standards and training. It appears likely that the success that the college graduate often experiences when placed in a supervisory position results not only from knowledge and skills but from an ability to work effectively with higher levels of police management personnel, many of whom are likely to be college trained. This type of acceptance by higher authority is soon recognized by subordinates, and usually it will make them more willing to accept the supervisor as it is felt that he or she has the full backing of the top brass in the department.

The role of the present supervisors in promoting the supervisory development program cannot be overemphasized. Officers who show promise as candidates for supervisory positions should be identified and given minor supervisory responsibilities as opportunities arise. Often, however, it proves undesirable to promote an officer to a supervisory position within the officer's own unit because it tends to put the officer in the untenable position of having to enforce discipline and to give orders to those who are "buddies." Thus, supervisors should be encouraged to identify talented personnel for possible assignment to other units in the department. This policy would encourage the accumulation of planned broad-based experience by means of rotating assignments. It is very desirable for a supervisor to have a broad base of experience.

While supervisors provide one of the most valuable sources of information about employees who have supervisory potential, the personnel unit of the agency should also help by screening the records of employees periodically for individuals who have been overlooked. Performance evaluation reports, test scores, education and training records, commendations, and other data contained in an individual's personnel file may reveal good candidates.

In conclusion, it can be said that a supervisory selection program should be tailored to the needs of the local organization, particularly when the department is large enough to afford such an approach. The tailored program has the advantage not only of meeting local needs, but also of being easily sold to the people who have to work with it. People tend to accept a program more readily if they know it was designed especially for their own organization, rather than borrowed from some other agency.

DISCUSSION QUESTIONS

1. What should be the aim of a good supervisory development program? How can this be accomplished?
2. What are some of the qualities of a good supervisor?
3. What is the drawback of using nomination by superior officers as the sole means of selecting supervisory personnel?
4. Name five procedures and considerations likely to be included in a good supervisory selection program.
5. Why should written examinations be validated within the local department?
6. What is meant by a *reliable* test? What is meant by a *valid* test?
7. What is meant by *item analysis* of a written test? What is meant by *factor analysis* of a written test?
8. Why might an oral board be part of the supervisory selection process?
9. How is the leaderless discussion group used to evaluate supervisory candidates?
10. What must be taken into consideration when weighing a candidate's performance ratings?
11. What is a peer rating? How can it be used in selecting supervisors?
12. In your opinion, is the trend toward substituting education for seniority as a requirement for promotion to supervisor beneficial or inadvisable?
13. When an officer from the ranks is promoted to supervisor, should this person be kept in the same unit or moved to a different unit? Why?
14. Discuss the role of present supervisors in developing new supervisory talent.

NOTES

1. William B. Melnicoe and John P. Peper, *Supervisory Personnel Development* (Sacramento, Calif.: California State Department of Education, 1965), p. 8.
2. Don Driggs and Paul M. Whisenand, "Assessment Centers: Situational Evaluation," *Journal of California Law Enforcement,* California Peace Officers' Association, Vol. 10, No. 4 (April, 1976), pp. 131-135.
3. Herbert J. Chruden and Arthur W. Sherman, Jr., *Personnel Management,* 2nd ed. (Cincinnati: South-Western Publishing Co., 1963), pp. 358-359.
4. Allen P. Bristow, "Survey Indicates California Law Enforcement Stresses Value of Education," *Journal of California Law Enforcement,* California Peace Officers' Association, Vol. 2, No. 1 (July, 1967), pp. 19-22.

5 The Psychological Aspects of Supervision

Inasmuch as a principal task of the police supervisor is the development of an ability to deal with other people, it is important for this individual to study some of the factors that influence both behavior and the roles that people play in their work environment. The supervisor must recognize the fact that things that are personally interesting may hold no interest at all for subordinates. The supervisor must understand his or her role and its relationship to the various roles assumed by subordinates under varying conditions. In short, a supervisor must have at least a basic understanding of psychology.

Aspects of Differences among People

People differ from one another in at least two respects. First, there are differences among groups of people—races, religions, social classes, nationalities, etc.[1] Second, individuals within the same general group may differ markedly from each other. For example, all police officers belong to the same general group, but no thinking person would say that all police officers are the same.

Some differences among people are highly desirable, whereas others are not. Society requires a diversity of talents, aptitudes, skills, and achievements. The scientist could not survive without the food grown by the farmer, and the farmer benefits by using methods developed by the scientist.

Pressures exerted by the community tend to push the individual toward conformity and uniformity. Nonconformity is desirable, within limits, in that it results in change, but when carried to extremes it often results in unconstructive, antisocial behavior. People who evaluate others are often influenced at the unconscious level by the person's conformance or nonconformance to group standards.

Basically, people are much more alike than they are different. Groups of people who have lived in an area for several generations adapt to their environment physically, mentally, and emotionally, and have probably developed unique customs and habits. However, they continue to resemble people in other parts of the world. The police supervisor must be aware of the basic similarity between people to understand fully the ways in which they differ.

There are several contradictory theories regarding human differences. One is that all people are created equal and that all handicaps and inequalities can be overcome by education.[2] By this theory there are no inherent individual limitations, and this encourages intolerance of the person who does not live up to his or her potential.

At the other extreme is the theory that differences are unalterable and that society must use the varied gifts of individuals to enrich the "common life." This viewpoint, too, rationalizes the existence of a privileged class and eases the conscience of those who prefer to ignore oppressed or underprivileged groups of people.

Between these two extremes is the moderate view which holds that no two individuals are identical and that each person should be considered unique.[3] This theory of individual differences considers that each person has a special pattern of aptitudes and interests, individual limitations, and an individual quality of intelligence. Science has proven that even one-celled organisms have individual patterns of behavior. If this is true of the simplest organism, then such differences are logically much greater in the complex human organism. The police supervisor must learn to accept and understand differences among people, then use this knowledge in a constructive manner.

Scientific study makes it possible, within broad limits, to predict behavior, to anticipate degrees of ability, and to determine certain personality characteristics based on an analysis of the group to which an individual belongs. However, it has been found that the differences among individuals within the same group usually are much greater than the differences among the averages of the groups as a whole. There is a good deal of overlap in the ranges of different groups being compared for the same trait. If you were to measure the intelligence scores of a thousand doctors, add them, and then divide the total by one thousand, the average score would undoubtedly be higher than that obtained for a group of a thousand laborers. However, an examination of the individual scores would undoubtedly show that some of the laborers were more intelligent than some of the doctors.

Regardless of whether any particular human trait can be attributed to environment or to heredity, it must be conceded that there are wide variations in human ability and personality. Some people are very intelligent, while others are extremely stupid; some have natural mechanical ability, while others are talented musicians.

Why are people different from one another? There has been scientific controversy about this question for a long time. One group places heavy emphasis upon the role of heredity and discounts the influences of environment, while the members of the opposing group attribute most differences to environment and discount heredity. Most contemporary schools of psychology believe that human beings do not inherit instincts or behavior functions *in toto*, but that every trait and reaction shown by the individual is dependent upon the interaction between heredity and environment. In other words, the effect that an environmental factor has upon the individual is influenced by his or her heredity. Conversely, the effect of all hereditary factors is influenced by environment. The interaction of the two types of factors is so subtle and complicated that it is nearly impossible to ascribe with certainty any particular trait or behavior pattern to one or the other.

The total effect of the interaction of genetic inheritance with environment can produce millions of separate and unique combinations of physical makeup. For this reason, no two people are ever exactly alike. Even identical twins who begin life with the same heredity will very shortly show personality differences because of the differences in their perceptions of the environment.

It appears safe to say that an individual begins life with certain tendencies and with a certain limit to his or her potential development in any given direction. But family, church, school, and ethnic group influence the individual continually, reinforcing some tendencies while ignoring others. A person may develop the maximum possible capability in one area, while another area is never developed at all. Some individuals develop maximum capability in several areas, while others never come close to realizing their potential in any area. The sum total of this interaction of environmental factors with heredity factors produces the highly unique human with individual personality, character, needs, wants, and drives.

The supervisor must be prepared to accept subordinates as they are—the end results of countless influences exerted over periods of many years. No individual can be molded into the ideal police officer by even the most highly skilled supervisor. The supervisor who attempts to mold subordinates in this manner is doomed to frustration and defeat.

The fact that the supervisor accepts subordinates as they are does not by any means rule out the possibility of helping them to adapt to the work situation. Many of an individual's work habits, attitudes, and personality traits are superficial enough to be altered. Through daily contacts, the supervisor helps subordinates move in the desired direction. The success or failure of the supervisor depends largely upon his or her success in handling human problems.

The job of classifying people becomes even more complex as we progress to a systematic study of human behavior. In the final analysis we learn that the only logical and accurate way to deal with human beings is to treat them as individuals. It is incumbent upon the supervisor to develop tools and techniques, as well as ways of thinking, for this sort of an approach to the problems of police supervision.

Leadership in the Context of Culture

The word "culture" generally refers to man-made environments and influences, unconsciously absorbed, that govern behavior and beliefs. In the broad sense, culture consists of knowledge, tradition, belief, art, custom, and law; it is the conventionalized behavior of society to which all persons more or less conform.

Artifacts are some of the physical manifestations of the culture of a society. In a police agency, which is a small society, the artifacts are items such as uniforms, weapons, insignias of rank, record forms, filing cabinets, restraining devices, and radio-equipped automobiles.

The culture of a society exerts very strong influences on people, but it must be realized that there are vast differences among cultures of various societies, and what is acceptable in one may not be accceptable in another. In some societies community bathing is the custom. Such behavior is considered nonstandard in our society and is subject to legal as well as moral sanction.

Culture largely influences job motivation and work habits. To undertake a study of any aspect of human behavior without attempting to understand the culture in which it operates is similar in effect to studying fish without realizing that fish live in water. To the extent that culture influences work habits, it is responsible for both good and bad performance. For example, the London police department, for many years, deliberately recruited young men from rural areas because it was felt that without sophistication and urban alliances they were more easily indoctrinated. Due to the lack of environmental bias, personnel from outside the district were integrated more rapidly in terms of behavior standards.

Police supervision and management are also influenced by the culture of the community. In most cases, a community has the sort of police department it wants and is willing to pay for—ranging from the corrupt and lawless at one extreme to the highly professional and efficient at the other.

Police supervision and management are also influenced by the culture within the agency itself. The size of the department determines the number of subcultures that develop within it. An example of this is the social stratification among detective, traffic, and patrol units of the large police agency.

There are two ways by which the supervisor can change the environment or culture of the work place. The most obvious way is to create a cultural change in the management situation by altering the social climate. This cannot be done merely by teaching the methods and techniques of human relations. Management must see the need to adapt to new methods and must be open to public feedback. It must then put those ideas to work in the management society and make that society a model for the operative society that it is trying to influence.

Human behavior can also be changed through education and training. Any person who has had exposure to education knows that it has had the effect of creating some personal change, even though the change may have been a minor one.

Supervisory Leadership Roles

The supervisor plays many roles in leading subordinates. While functioning as a *climate-setter,* the supervisor sets the patterns of interaction for subordinates, influencing them by example. If the supervisor is moody and uncommunicative, subordinates will quite likely imitate this behavior. Conversely, if the supervisor is cheerful and outgoing, these traits will be communicated to subordinates and will be reflected in their attitudes.

The supervisor also functions as a *symbol*—the standard-bearer and rallying point of the department. In the minds of subordinates and the public the supervisor is the embodiment of the department. How often have you heard a citizen demand to "speak to the sergeant"? However, the true leader will resist the tendency to become a figurehead and will not abandon other leadership responsibilities in order to satisfy a public image.

Another role of the supervisor is that of the *objective-setter.* This is the more or less traditional role of every leader—deciding where the group is going and how it will get there. The statement of objectives is one of the more sensitive aspects of leadership. The supervisor who takes too little initiative in determining direction runs the serious risk of being led by subordinates. On the other hand, if goals are set that are too much at variance with the goals of subordinates, the supervisor may find no followers. The alert leader will keep in close touch with the group so that the proper balance can be achieved between too much and too little goal setting.

The supervisor also functions as a *protector.* This role may be more obvious in the military and political arenas, but it is still of great significance in the police service. Its importance varies with changing conditions; the individual or group that feels threatened will consider protection to be the most important role of the leader, while the secure individual or group will prefer to have the leader perform other functions. A subordinate needs the security of knowing that the supervisor is going to be totally supportive when the occasion demands. The police supervisor who backs down under pressure will soon lose the respect of subordinates.

The supervisor is a *conciliator.* In almost any group, there is bound to be some internal dissension from time to time, and one of the important tasks of the supervisor is to see that differences are resolved fairly. In this function the supervisor acts in a judicial capacity—arbitrating, trying to see the larger issues, and attempting to gain mutual understanding between the warring factions. The need for the supervisor to assume this duty varies according to the climate in the group—a climate which the supervisor is very influential in setting.

The supervisor functions as the *ego-ideal* of the group. Many adults tend to be somewhat embarrassed by any implications of hero worship. It is perfectly acceptable for a child to have heroes, but adults do not like to accept the fact that they need heroes too. If the hero can also be the supervisor, the individual and the supervisor are both quite fortunate. A distinction needs to be made between a child's hero worship and an adult's healthy identification with a leader. An adult's identification with another person is much more discriminating, and it is usually something less than 100 percent admiration.

The "hero worship" aspect of leadership can function in reverse as well; some subordinates, unable to identify with a good supervisor, will resent his or her direction. To one individual the supervisor may be a hero, to another, the enemy. The supervisor who notices a negative reaction from a subordinate can, if he or she prevents an emotional, personal response, find ways to utilize and channel these feelings in constructive ways.

The supervisor acts as a *decision-maker*. This role is one of varying importance, depending largely upon personalities and departmental practices with regard to the delegation of authority. When the supervisor has authority to make decisions, subordinates prefer that certain criteria be met. The supervisor who meets the following three criteria in the eyes of subordinates is well on the way to being recognized as a good leader.

First, with regard to errors, the supervisor should have a pretty good batting average, one that is well above chance but humanly below perfection. If the supervisor makes too many mistakes, subordinates will become anxious and lose confidence in that person. On the other hand, if the supervisor makes a mistake occasionally, it shows a natural human frailty. The infallible supervisor is a difficult person to live with.

The second criterion is a sense of appropriate timing. Subordinates will judge a supervisor by the person's willingness to make a decision at the time it should be made. While snap decisions may be unpopular, they are generally preferred over procrastination and indecisiveness. (This is particularly true in field supervision in the police agency.)

The third criterion is accountability—the willingness of the supervisor to accept the consequences of decision making, including blame when a bad decision has been made.

The supervisor functions as a *communicator*. There are many successful leaders who are poor communicators by formal standards. They cannot deliver a speech, do not listen well, write poorly, and cannot read well. From the psychological viewpoint, however, the important consideration is not whether the supervisor communicates well by official or academic standards, but whether the supervisor and subordinates understand one another. On occasion, this sort of communication can be quite informal. Subordinates may learn to interpret the supervisor's grunts or the way an ear is scratched or the way a pipe is smoked. An important aspect of successful communication is the belief by subordinates that the supervisor intends to communicate with them. If the supervisor seems willing to communicate, subordinates are usually willing to overlook a particular mode of expression. While clear communications are a necessity for a supervisor, it must be recognized that there are many successful styles of communication. What is important is that the style be suited to the needs of leader, group, and situation and that there be little or no misunderstanding.

The supervisor functions as a *disciplinarian*. Discipline is often defined as teaching that molds, corrects, or perfects. A basic supervisory function is to provide a climate in which subordinates can learn. Discipline ranges from teaching by example and giving encouragement to drastic punishment. The basic purpose of discipline is to aid the subordinate in the achievement of desirable goals. Disciplinary practices vary widely as different supervisors desire different goals and use various methods to attain them.

The supervisor is a *subordinate*. All police supervisors are also subordinates, whether of a lieutenant or of the chief of the agency. In part, a supervisor's success as a leader will depend upon the relationship achieved with superiors. If a supervisor wishes to be effective, he or she must demonstrate the ability to be a successful follower.

The supervisor is also a *peer*. In a police agency the supervisor usually has peers—sergeants, lieutenants, captains, or others whose rank is equal. Relationships with subordinates are usually affected by the relationships established with peers. The supervisor who has succeeded in building a relationship of mutual confidence and trust with other supervisors is most likely to be successful in the leadership of subordinates.

The supervisor functions as a *role-clarifier*. One of the most important duties of the supervisor is to help subordinates identify and clarify their roles and make it possible for them to fill their respective roles comfortably. This is achieved by working with subordinates on understanding the breadth and requirements of their jobs in the department, by helping them resolve incompatible aspects of their roles, and by encouraging changes in roles as tasks and individuals grow.

Everyone performs a variety of jobs simultaneously in everyday life. A police supervisor, for example, may also be a husband, wife, father, mother, a leader in the church, a youth leader, the president of the police association, or a boating enthusiast, as well as a subordinate to his or her boss.

A supervisor's effectiveness and happiness depend to a large degree on how clearly these roles are perceived by the individual as well as by his or her associates. And roles may change. An individual transferred from the patrol division to the detective division will find that old roles must be abandoned and new ones assumed.

Clarification of one's roles is necessary, but the act of clarifying is not by itself sufficient for effective performance. An individual's roles must also be compatible. A police lieutenant can be simultaneously a supervisor, informal leader of other supervisors at the same level, and subordinate to a captain, because these roles fit easily within the structure of the department. But suppose the lieutenant is also a member of a minority ethnic group that frequently clashes with the police in the community. This role may then be incompatible with the other roles the lieutenant plays as a member of the police department. Associates may feel that the lieutenant acts too much like a member of the minority ethnic group, while members of the minority group may feel that the individual acts too much like a police officer. Meanwhile, the lieutenant is torn between both roles.

It is the job of the lieutenant's supervisor to help this individual. First, the supervisor must objectively evaluate the situation. Are the minority group's stated goals such that conflict with the police is inevitable, or are certain members of the group forcing the conflict? Or could it be that the group was formed in response to what was felt to be unjust police pressure on members of the minority community? An answer to this question will provide the first clue as to how the situation should be handled. It may well be that the lieutenant who is filling this dual role is one of the most valuable individuals on the force. As an "insider" the lieutenant may act as a moderating force, toning down the more radical members of the group. This person may be the only effective communications link between the group and the police, bringing the feelings of the community to the attention of the police administration. In either of these cases, the lieutenant's supervisor should do everything possible to make the lieutenant more comfortable in this dual role. It might help to change the person's assignment, or to give the person a chance to talk to subordinates in a body and explain just what the minority group is all about. Of course, it might be that roles are intrinsically incompatible and, in that case, the individual might have to make a choice between loyalty to job and loyalty to background.

The supervisor also acts as a *sponsor*. When functioning as role-clarifier, the supervisor helps subordinates find themselves and their niches. When acting as a sponsor, the supervisor encourages subordinates to strive and to expand. Under this sponsorship they grow as much as their capabilities will allow. The good supervisor assists subordinates in their personal growth because, as they become more proficient, their effectiveness in the department also increases.

As we have seen, there are many kinds of supervisory roles in police management. Because leadership is practiced within such diverse patterns, there is no single mold into which the supervisor must fit. The department should encourage a reasonable degree of flexibility and individuality so that the supervisor can prepare for as many leadership roles as he or she is capable of filling.

A realistic supervisory development program recognizes the fact that leadership training is a continuous process that is best taught by example.

The character of leadership is not simply a combination of desirable traits, but is a complex interaction between the supervisor, subordinates, and the demands of the situation.

Often the supervisor is called upon to serve as *counselor* to subordinates. While obviously not a trained clinical psychologist, a wise supervisor will know and recognize the symptoms of the various types of personal problems that are common among men and women frequently subjected to great stress, such as occurs in the police service.

The Supervisor and the Problem Drinker

Alcoholism is a good example of what can result from unresolved personal problems and unrelieved tensions. Business and industry are beginning to take a new look at the old problem of alcoholism among employees;[4] perhaps it is time for police agencies also to face this problem realistically. A large number of private employers are discovering that a highly significant percentage—if not the majority—of problem drinkers can be rehabilitated.

Although the police supervisor cannot be concerned with the causes of alcoholism, a concern must exist for the effects. Alcoholism, regardless of its cause, results in increased use of sick leave, disciplinary problems, errors in judgment, accidents, poor public relations, and a host of other problems. Because the label "alcoholism" implies a specific diagnosis, the term "problem drinking" is probably more accurate in describing the work problems that confront the supervisor. Generally, the police supervisor avoids diagnosing or indicating to the subordinate that a drinking problem exists unless there are some definite overt acts and unsatisfactory work performance relating to the use of intoxicants.

There are some cues that the alert supervisor can pick up which may suggest that an employee has a drinking problem. The California State Personnel Board suggests that the following are indicative of a possible drinking problem:[5]

    1. Telephone calls or letters from creditors of an employee who is a "heavy" drinker.

    2. Any instance of missing work because of intoxication, and/or hangover, particularly just before or just after days off duty.

THE PSYCHOLOGICAL ASPECTS OF SUPERVISION

3. Any on-the-job injury due, in part, to drinking.

4. Any off-the-job injury due to drinking which causes loss of work time.

5. Significantly lower quantity or quality of work than previously performed by an officer who is a "heavy" drinker. (The supervisor should examine this problem regardless of whether or not the man drinks.)

6. Any off-duty arrest or report involving drinking.

7. Any instance of an employee indicating that drinking is a problem.

8. Any physical or mental problem to which alcohol is significantly contributory which affects job performance.

9. Any instance of drinking or intoxication during duty hours.

10. Employee reporting for duty with a noticeable odor of alcohol on the breath.

Most successful employer-sponsored programs to rehabilitate problem drinkers handle the disciplinary problems of these employees in the same manner that they would handle comparable problems of an employee who was not a problem drinker. However, it is not reasonable to assume that a disciplinary problem can be resolved without resolving the drinking problem. The authors recommend the following programs for handling problem drinking, assuming of course that such programs are acceptable within local departmental policies.

At the first instance of unsatisfactory work performance the supervisor should warn the subordinate that job performance is below standard, that this will not be tolerated, and that the situation will be handled in accordance with normal disciplinary procedures. The individual should be advised to seek help if problem drinking is the cause of this difficulty and should be told that seeking such help will not place the person's job in jeopardy. Recommended sources of professional assistance should be made available to the employee.

If there is a second instance of unsatisfactory work, the supervisor should recommend a short suspension of from one to seven days, depending on the nature of the incident, and again inform the individual that this behavior is not acceptable and that more severe penalties will result if it is not corrected. At this time the employee should again be given the names of persons or organizations to contact for assistance.

If a third incident occurs, the supervisor should recommend a longer suspension from duty of from ten to thirty days, depending on the incident, and the individual should now be informed that dismissal action will be initiated if the problem recurs. Once more, the employee should be strongly encouraged to request and accept assistance in solving the problem.

If a fourth incident should occur, the supervisor should recommend that the employee be dismissed from the department.

The Supervisor and the Mentally Ill Police Officer

Occasionally, despite the most rigorous form of psychological and psychiatric screening, and despite information obtained from a detailed background investigation, a person will be hired who is psychologically and emotionally unsuited for police work. It is an extremely important responsibility of the supervisor to be alert to this possibility and to take appropriate measures to separate such persons from the police service—preferably early in the probationary period. The supervisor must to be able to recognize the symptoms of actual mental illness as well as those of emotional instability; either of these can reduce a subordinate's effectiveness in police work. If an individual is found to be unsuited to the police job, it is a disservice to the department, the public, and the individual to continue employment.

An example is found in the experience of one of the authors, who noticed that a new man displayed an abnormal fear of firearms during range practice. When the fear continued over a prolonged period, it was recommended that the man be dismissed, but a higher authority decided to allow him to continue through his probationary period. The first night the man was on duty in a patrol car, he encountered an armed-robbery suspect and was shot and killed. Witnesses later stated that the officer had seemed hesitant about drawing his weapon. It would appear that fear and hesitancy cost the officer his life.

The good supervisor should be able to distinguish between the overly authoritative and possibly emotionally unsuitable officer and the normally eager and, in some cases, overzealous new recruit. It is a difficult task for the supervisor to keep a tight rein on subordinates so that they remain within the bounds prescribed by law and by the department's policies, and at the same time allow enough freedom so that their initiative and enthusiasm are not stifled. However, it is a challenge which must be met if the supervisor is to be effective and his or her unit is to operate at peak efficiency.

It is never possible for a supervisor, or anyone else, to be aware of all of the personality traits of any individual; however, it is possible to observe the relations between a subordinate and the public, fellow officers, and supervisor. The majority of these contacts should be positive in nature, although allowance should be made for the occasional off-day we all have now and then.

Cases of persons suffering from severe mental illness in the form of psychosis are relatively rare among police personnel; when they do exist, it is the responsibility of the supervisor to identify them as quickly as possible and to refer such people to a professional for help. It will be probably necessary to recommend termination of employment. The most difficult type of psychotic to detect, and also the most dangerous, is the paranoid or paranoid type of schizophrenic whose well-organized ability to delude allows this individual to conceal symptoms and appear relatively normal. However, an alert supervisor may be able to observe feelings of persecution or expressed antagonism in such a person toward individuals or groups who are "no good" or "out to get" the individual.

The average supervisor is not sufficiently trained or experienced to handle such cases. A supervisor has a responsibility for detecting mental illness, but the responsibility stops with identification.[6] The supervisor should be gentle and reassuring in dealing with such an individual, but no more than this can be demanded from a nonclinician.

Neurotics are common in nearly every work group. An individual who has no trace of neurosis in his or her personality makeup is probably also a "vegetable." The degree of neurosis and the adjustment made to it by the individual are the keys to whether or not he or she functions well as a person. It is when the neurosis moves toward its more extreme forms that it becomes a real problem for the supervisor. The most obvious neuroses are the various forms of hysterical reaction and acute depressions with suicidal tendencies. Hysterical persons tend to have physical illnesses that have no apparent organic origin, while those subject to acute depressions are likely to express feelings of unworthiness or to speak of "ending it all." The supervisor should make every effort to get these people to seek professional help and should feel most definitely responsible for the prevention of suicide. It might be noted here that the statistics over the years indicate that police personnel as a group have a higher suicide rate than do other occupational groups.

Beware of the individual who expresses hatred for any person or group of people. This is a major signal that should alert any supervisor. The individual may be simply a minor troublemaker who will confine behavior to needling other employees or playing practical jokes. On the other hand, this person may needlessly beat or kill someone. Such individuals should be watched for any signs of latent viciousness in their activity.

In summary, it is the job of the supervisor to regard all subordinates as individuals and to create a work climate that is conducive to a happy and efficient unit. The supervisor must weed out the misfits, must offer all assistance possible to subordinates, and must also be alert to emerging problems, taking appropriate action before the situation becomes acute and disrupts or endangers the unit.

DISCUSSION QUESTIONS

1. Why are individual differences necessary to society?
2. In what ways are all police officers alike?
3. In what ways are all police officers different?
4. In what way is individual development limited by heredity?
5. What is the role of environment in individual development?
6. What are the implications of the "management culture" for the police supervisor?
7. Why is a change in the "social climate" within the police agency likely to effect a change in the agency's culture?
8. What means can a supervisor use to bring about changes in the behavior and attitudes of subordinates?
9. Of what significance are the various roles played by the supervisor in his or her daily contacts with subordinates?
10. Why do different employees view the supervisor in different ways?
11. What steps can a supervisor take to handle the problem of a drinking subordinate?
12. How is the problem drinker likely to be affected in terms of work performance?
13. What should be done about the "badge-happy" recruit?
14. What is the supervisor's responsibility in regard to severe mental illness in a subordinate?

NOTES

1. John M. Pfiffner and Marshall Fels, *The Supervision of Personnel,* 3rd ed. (Englewood Cliffs, N.J.: Prentice-Hall, Inc., 1964), p. 49.
2. Leona E. Tyler, *The Psychology of Human Differences,* rev. ed. (New York: Appleton-Century-Crofts, Inc., 1956).
3. Pfiffner and Fels, op. cit., p. 50.
4. "Alcoholism—A Problem of Increasing Concern," *Management Manpower Bulletin,* California State Personnel Board Bulletin No. 3 (August, 1967), p. 1.
5. Ibid., p. 2.
6. Paul B. Weston, *Supervision in the Administration of Justice* (Springfield, Ill.: Charles C. Thomas, 1965), p. 64.

6 Motivation

Motivating Subordinates to Work

Motivating individuals to work is one of the most difficult problems confronting the police supervisor. Human beings present a broad spectrum of motivational problems that often defies analysis. However, certain common methods of motivation seem to work in most cases. These methods are included in the following presentation. They are not guaranteed to work in all cases, but they should prove effective most of the time.

Principles of Motivation

An analysis of the methods employed by hundreds of executives in industry has shown that there are seven basic principles of motivation utilized by successful leaders that can be applied easily to police personnel.

Encourage Self-Involvement

People work more effectively on a job they want to do and that they feel is theirs than they do in a job they feel someone else wants them to do. Just as people invest themselves more intensely in jobs they like, they also prefer to work in an area where they feel competent. Three individuals doing the same job will handle it differently, according to their own experiences and aptitudes. A lack of involvement results in a psychological withdrawal from the job.

Subordinates must be trained before they can be properly motivated; simply telling people to do a job will not make them perform effectively.

They must first accept what they are told. The example set by their supervisor and the way in which the goals of the organization are presented will particularly determine whether or not the employee will do the job enthusiastically and well. People who perform in a perfunctory manner often claim that the only reason for carrying out an assignment is that "somebody higher up wants it done."

A supervisor can encourage self-involvement by having a subordinate devote a fair amount of time—as much as possible—to jobs he or she likes and does well. At the same time, a subordinate should be helped to become competent in weak areas and to develop tolerance for those tasks he or she dislikes. When practical, the supervisor should make suggestions and recommendations rather than give direct orders. The seeds of the idea should be planted so that a subordinate uses it as his or hers rather than as the supervisor's idea. If a subordinate is allowed to criticize certain routine assignments, even though no change can be made in the plan, it will help to eliminate the feeling that something has been forced upon the individual. The following are some of the methods that can be utilized by supervisors to make police officers more proficient in weak areas and to broaden their experience.

1. Rotation of patrol beats and working hours

2. Assignment to specialized units, e.g., traffic, juvenile, investigation, training, and others of an unusual nature

3. On-the-job training under supervision

4. Encouragement to continue formal education

5. Role-playing which approximates real conditions

A supervisor should let a subordinate know the outcome of work assigned. The subordinate may be doing a small part of a large project and may naturally be anxious to know the final result. When so informed, a person can be motivated without realizing it. A police supervisor often has to order a person to do a specific job in a specific way and, in many instances, is not able to relay to the subordinate the outcome of this work. Expediency will have to govern the use of this suggestion. When detectives let an officer know the result of a field arrest, it certainly makes the officer feel part of the team. This method is accepted practice in many departments.

Give Freedom but Keep Control

Delegation of responsibility inspires a feeling of self-confidence in a subordinate and makes the individual feel that the supervisor has faith in him or her. It gives the subordinate a chance to learn, to show initiative, and to make a personal contribution.

New supervisors will often "jump" calls rather than let their officers do the job. They may be unable to break away from a former role, or they may be attempting to show by example that they are good police officers, believing that they are gaining the confidence of their subordinates in this way. But the subordinates are offended and lose their sense of initiative and responsibility. Soon they will feel that they should let the supervisor handle it, as he or she is going to do this anyway.

Some supervisors will not give their personnel calculated opportunities to make mistakes or to use initiative. If they think an officer is about to make a minor error, they will jump in and give advice on the proper procedure. Subordinates normally resent this control and are demoralized by it. A supervisor who uses the calculated-opportunity theory in situations that will not cause embarrassment to the department or jeopardize the effectiveness of a police investigation will discover the value of this procedure in producing greater self-confidence in subordinates.

Some supervisors feel needed only when subordinates are dependent upon them. This dependency is dangerous because it increases the chances that subordinates will make poor decisions under pressure when the supervisor is not there to provide the solution. How many supervisors have discovered that some subordinates "turn off their brains" after calling a supervisor to the scene of an incident? They appear just to sit and wait. A subordinate should be encouraged to respond to the supervisor who asks "What is your plan?" The officer should be taught to consider the law and the agency's policy, to use good judgment, and to provide alternatives when developing a plan.

Delegation of responsibility is difficult for many supervisors; they are afraid that, as a result of such delegation, something might happen to jeopardize their position. Such supervisors are probably too cautious and may be underestimating their subordinates. In some cases the supervisor's need for power may be too great and the supervisor may see delegation as a lack of control. But handled correctly, control can still be maintained. A good supervisor will delegate responsibility only after setting up controls which enable the individual to take corrective action if things go wrong. Four factors are important in delegation:

1. The subordinate must be properly trained.

2. The subordinate must be given responsibility in a step-by-step manner.

3. The subordinate's mistakes must be corrected and successes recognized as he or she moves along.

4. The delegator must have controls set up so that he or she can move at any juncture to prevent action that would seriously jeopardize the delegator's future or that of the subordinate.

Help Subordinates Identify with Others

Positive motivation occurs when a subordinate feels that he or she is not just another face lost in the crowd, but is valued by supervisors for unique qualities as an individual. The supervisor should take a sincere interest in the personal fortunes of subordinates.

A supervisor should develop a general relationship with subordinates wherein they discuss areas of mutual interest such as children, sports, entertainment media, and building projects. Discussions of anecdotes, jokes, and other nonpolice subjects should not be discouraged, because they will lead eventually to constructive suggestions and will aid in the motivation of subordinates toward the organization's goal.

Subordinates will work hardest when they like their supervisor and feel that the supervisor likes them. A good basis for relations is the middle ground between aloofness and disinterest on the one hand and overfamiliarity on the other; either extreme can be disastrous for the police supervisor.

Give Credit

Psychological studies indicate that recognition is one of the most significant factors contributing to the motivation of individuals, and that it is of more importance than responsibility, salary, advancement, or the work itself. Recognition has both positive and negative aspects. Praise and criticism are opposite sides of the same coin, but must be considered separately. Recognition not only gives a subordinate a feeling of appreciation, but also lets the employee know how well he or she is doing. Most individuals are, to some degree, anxious about their performance. When anxiety is high, subordinates need frequent praise and intermittent reassurance to allay fears. When the anxiety level is low, subordinates do not require constant praise, as most people will assume that their supervisor will tell them when they are not doing a good job. As a motivational technique, the value of praise is directly related to the importance of ego in the personality makeup of the normal individual—and the ego-gratifying needs of people are enormous. Who is not affected by flattery? Praise, even when we doubt its sincerity, is exhilarating.

Some supervisors feel that praising their subordinates will cause the subordinates to relax their performance, but the opposite is usually true. People react best when they receive recognition for their good work. Commendations given in public are most effective, because the entire organization knows that good work is being recognized. On the other side of the coin is condemnation or the rebuke which, when required, should be administered in private. The officer's prestige and status in the peer group will be lowered if the officer is reprimanded in public, and he or she might withdraw into a shell or work against the organization. The modern police officer is intelligent and aware of good supervisory techniques; supervisors who do not use effective methods may lose the confidence and respect of their subordinates.

Two types of supervisors fail in the area of motivation; each type makes the same mistakes, but for different reasons. The first type is highly self-sufficient and confident, with so much self-assurance that he or she needs little from the outside. Assuming that other people are equally constituted—or that they should be—this person finds no need to give praise. The second type is the guilt-ridden individual who feels that he or she rarely comes up to personal standards. Because such a supervisor seldom feels that he or she has done a good enough job, this person seldom thinks subordinates have either; hence, they do not deserve a compliment. With either type of supervisor, subordinates do not get the credit they deserve, and morale suffers.

It has been stated previously that credit for good work must be given publicly, and that reprimands for mistakes or inferior work should be pointed out in private. A good supervisor will solicit suggestions and ideas and will give recognition to those who offer them. If a suggestion is used or adopted and is effective, the supervisor will give due credit. If a suggestion is adopted but is unsuccessful, the supervisor will not blame the originator. The effective supervisor knows the value of communication and will notify the subordinate of the action taken on the subordinate's suggestion. If it was not accepted, the supervisor will explain why it wasn't. It is imperative that supervisors do not steal others' suggestions or ideas and pass them on as their own.

Show Confidence

A prime requisite of leadership is self-confidence. Subordinates need to have faith in their superiors and need a sense of direction and purpose. They also need someone who can make vital decisions and to whom they can turn for assistance and guidance. Confidence breeds confidence; a real leader may often give subordinates courage that they otherwise would not have.

There is a significant difference between confidence that is openly manifested and that which is not. Strong leaders appear manifestly confident; regardless of the real level of their confidence, they appear sure of themselves. Even a supervisor with serious doubts about personal ability may nevertheless present an outward picture of calm assurance that enables the individual to assume the role of leadership. Conversely, the person who feels confident inwardly may not be able to project or communicate this conviction, and therefore makes a poor leader. The supervisor's confidence should be great enough to override the fears and doubts of subordinates and to give them a strong sense of security.

Assign Blame

An employee needs to know whether or not he or she is going in the right direction and relies largely on the immediate superior for this sort of information. Positive feedback in the form of recognition, praise, and reward reinforces the employee's drive to continue in the proper direction. Negative feedback—or criticism, correction, and punishment—teaches the individual what *not* to do. Thus, both credit and blame are important as motivational techniques. Subordinates need to know that they will be told of specific mistakes and that they will be informed if their general performance falls below the department's standards. If they know where they stand with the supervisor, they will have the confidence necessary for the type of decisive action so vital in the police service.

From the psychological point of view, most people expect to be reprimanded when they have done something wrong. Properly administered criticism will help relieve their guilt feelings. However, some officers are not quite so conscientious, and they are less prone to feel guilty when they make a mistake. But if these individuals are not criticized, they are likely to exploit what they consider a weakness on the part of their superior.

Instill Fear

While it is not commonly thought of as a leadership attribute, the ability to instill fear can play an important part in specific cases and is a definite factor in motivation. Fear of bodily harm, such as that which was utilized by old-time Army and Marine Corps drill instructors, is not the most effective weapon. The instilling of fear should be refined and subtle, and more psychological than physical. A scowl of displeasure, without any words, can sometimes be highly effective in motivating a subordinate to work harder.

Generally, any successful motivator generates a certain amount of apprehension. A good supervisor must be tough and demanding. Subordinates must be certain that the supervisor has not only the authority, but also the courage to discipline them. Yes, courage! It is natural to try to avoid unpleasant situations. A supervisor who ignores or tolerates infractions of the rules takes the much easier path. Many supervisors feel that they will become unpopular if they reprimand or discipline subordinates, and there is strong motivation to "belong" and to resist action that might cause subordinates or peers to disapprove. Even though it is unpleasant, discipline must be maintained. The techniques used to maintain discipline must be employed in a manner that will not disrupt the organization.

Positive and Negative Motivation

The seven basic principles of motivation discussed in this chapter will have the most positive results for the supervisor who knows and understands the deep emotional responses subordinates have to their jobs. These feelings may drive and motivate the individual positively, negatively, or perhaps both at the same time, but they are always present. In the police profession these emotions become acute, intense, and often very visible due to the inordinate tension and pressure of the job.

Some of the positive motivators are reflected in an individual's desire for personal achievement, dignity, and independence, or in the continual struggle to meet or exceed personal goals and be suitably rewarded for the effort. Men and women are motivated by a creative environment where personal worth is continually affirmed and where competence is appreciated. People are motivated positively as well by a continuing challenge, by the trust of others, by intellectually stimulating work requirements, and by the simple feeling of being "needed" and productive.

On the negative side, a person may be motivated in unhealthy and nonproductive ways by feelings that he or she is unable to communicate, or by work conditions which are intolerable or unfair in the individual's eyes. Examples of these inverse motivators may be feelings of personal insecurity, feelings that supervisors have little or no interest in the subordinate, and feelings that unjust discipline and undue favoritism are being practiced by unfriendly and incompetent supervisors. These and many other types of negative feelings often lead an employee to perform at minimum capacity. Such feelings may also result in the employee's general dismay with departmental leadership and policies, and in great personal frustration for the individual man or woman.

Police Supervisor's Motivation Checklist

1. Don't tell yourself glibly that you rank reasonably well in all categories discussed in this chapter, but instead give some serious consideration to each point. Ask yourself what actions on the part of your superior either increase or decrease your motivation, and what are the differences between the way your superior treats you and the way you treat your subordinates.

2. Study the concept of self-involvement and check to see how well you apply it.

3. Find the leadership approach that fits your personality and is most natural for you. Capitalize on those tendencies that help you to be an effective leader, and attempt to limit or modify those characteristics that do not.

4. Make a concerted day-to-day effort to become more sensitive to the needs of your subordinates and to find out what motivates them. In doing so, guard against the natural tendency to expect other people to be like yourself. They are individual and different.

Properly utilized, these principles of motivation will increase work and give a supervisor subordinates who will assume greater responsibility, work with more enthusiasm, correct their own deficiencies, use more initiative, follow directions better, and attain goals more effectively.

DISCUSSION QUESTIONS

1. How does a supervisor get a subordinate "involved" in performing a disagreeable job?
2. In what ways is recognition important to subordinates?
3. What forms might recognition take in the police service?
4. Why is self-confidence so important to the supervisor?
5. Why is fear a factor in motivation?

7 Leadership

Former President Eisenhower once stated, "Leadership is the art of getting somebody else to do something you want done because he wants to do it." True leadership is a positive rather than a negative force and is based on cooperation and mutual trust, not coercion and fear. The essence of real leadership is the ability to obtain from each subordinate the highest quality service that he or she has the ability to render. Good leadership *does not* involve the exercise of authority through commands coupled with threats of punishment for noncompliance, nor is it related to the strength with which commands are barked. No organization can function efficiently without competent leaders, but this is particularly true in a police agency, because the independent nature of police work makes it possible for an officer either to follow or not to follow the leader.

Leadership is the ability to influence people to cooperate in the achievement of some common goal in such a way as to command their obedience, confidence, and loyalty. A leader must be able to understand human emotions and to handle each person individually. Anyone can *boss,* but the supervisor who has personnel working *with,* rather than *under* or *for* him or her is the true leader of people. Despite some opinions to the contrary, few persons are born leaders; the ability to be a successful leader is developed. The development of leadership ability depends largely upon the supervisor's ambition, determination, and desire to be a true leader—not a boss, and not a "supervisor."

Certain characteristics mark a leader of people. A leader enjoys seeing others develop under his or her guidance and direction, but never makes this pleasure obvious. A leader is direct and forthright in all actions and possesses fairness, moral courage, humanity, and a sense of humor. In addition, a leader has a sense of devotion to duty and a realism that allows the individual to see things as they are, not merely to indulge in wishful thinking. The leader understands human nature—its frailties and peculiarities—and is never disillusioned by the acts of others. Rather, the leader takes people as they are and directs their interest, cooperation, and effort to a common cause.

Improving the Ability to Lead

Supervisors sometimes fail to observe all the rules of good leadership, because no one can always remember all of them, and no one has enough self-control to avoid all mistakes. It can be shown that different successful leaders use quite different methods; however, this does not prove that they are born leaders, but rather, that they have adopted different methods to suit their personalities. Leadership is a learned skill. Some leaders succeed in spite of obvious faults, because they are outstanding in other ways. They could be better leaders if the faults were eliminated.

For example, one male supervisor may be impatient and extremely critical, but he gets good results because he is decisive—he gives orders and answers promptly—and he freely and openly admits his mistakes. He appears to be disliked, but is respected. Subordinates tolerate his shortcomings because they have confidence in him.

No one is, or can be, a perfect supervisor, but all leaders can become better administrators by following this program of improvement:

> 1. Become conscious of as many shortcomings as possible, and try honestly to correct each of them.
>
> 2. Take full advantage of every opportunity to use good leadership tactics.
>
> 3. Develop a reputation for possessing leadership characteristics to an outstanding degree by practicing them constantly.

The good practices will help counteract the mistakes. Time and effort are required, but good practices are easy to follow. The result will be better performance from subordinates and better job satisfaction from both supervisor and subordinates. Leadership is established by winning the respect, the confidence, and the loyalty of subordinates. In a sense, the leader must be a salesperson, for subordinates must be "sold" on their supervisor.

How to Win Respect

The supervisor must have the respect of subordinates, for men and women will not give loyalty to, or have confidence in, a person they cannot respect.

Lack of respect for a supervisor is followed by a lack of respect for instructions.

Encourage Free Speech

No one consciously wants subordinates to be "yes" people. But too many supervisors encourage or require their subordinates to be "yes" people by a negative attitude toward objection and criticism. This attitude causes the supervisor to lose the benefit of independent ideas, information on operations that require attention, correction, or improvement, as well as the respect of subordinates. The supervisor should welcome suggestions and criticism, consider them, and, if valid, take corrective action. If suggestions or criticism are not offered, they should be sought.

Finish What Is Started

Inaugurating a new plan or campaign involves: (1) adding to the work of subordinates, (2) attempting to improve efficiency, and (3) putting a plan and instructions to a test. Starting a program and failing to follow through loses the time, effort, and energy spent developing it and permits subordinates to disregard instructions. It also may encourage them to disregard other current orders and to pay less attention to future orders. Whatever is started should be kept track of and continued as long as there is a reason for it. If a better system is developed or the need for the activity ceases, it should be cancelled. When a project is cancelled, the cancellation should be announced or published with an explanation. In this way, the employees will know what is going on and will respect the supervisor who has been considerate of them.

Know What Is Going On

The supervisor must know what is going on, what kind of a job is being done by subordinates, what job conditions are like, what forthcoming changes may affect the unit, what irregular or improper habits or practices may be developing among the personnel, and what changes in the operating procedures of related agencies may affect the unit.

Statements, actions, or decisions that show the supervisor lacks knowledge diminish the respect of subordinates. This is particularly true when the information is common knowledge, and "everyone knows but the boss."

A supervisor's lack of knowledge makes it possible for subordinates to fool him or her by falsifying reports, working short hours, taking excessive time on breaks or personal business, and misusing the department's supplies or equipment.

In order to know what is going on, it is necessary to keep up to date on bulletins, orders, and statistical analyses, and to spend time with subordinates. The supervisor should ask questions, engage in casual conversation, and be alert for leads which will require follow-up. One cardinal rule in this respect is never to violate the confidence of a person who gives information. If a subordinate is embarrassed, harassed, or criticized because the supervisor identified or quoted the subordinate, there will be no further information—from this person or anyone else. The supervisor should also be alert for information or leads from citizens, business leaders, officials of related agencies, and others. The supervisor should never seek to "save face" by pretending knowledge of affairs and brushing off further discussion. Information or intelligence concerning the unit is not considered squealing or snitching; it is essential to good supervision. Both employees and the department benefit from a system which brings difficulties to the attention of supervisors. This is particularly true when conflict is resolved before dissatisfaction festers and spreads, or before misconduct damages the good name of the department.

Avoid Unnecessary Activity

Only work or activity that is really necessary and worth the time, effort, and expense it will require should be requested. Few things are as demoralizing as having to do work which is obviously of little value. Unnecessary instructions should also be avoided; they arouse resentment, humiliate the subordinate, and imply the supervisor's lack of confidence in his or her personnel.

Expect Good Work and Conduct

Subordinates must know that their supervisor has confidence in them, and that it is expected they will do their best. Work which is of poor quality should not be accepted. By accepting a poor job, the supervisor is approving it, and correction becomes much more difficult. Tact and patience may be required to show subordinates how the job should be done, but improvement will not occur otherwise. Standards should be set as high as possible and be worked toward. It is the little errors—the small omissions and insignificant oversights—that must be watched for. Together, these errors add up to the poor job. Men and women tend to perform according to what is expected of them. If standards are high and substandard work is returned for correction, the overall quality of work will be high. Adequate supervision restricts the growth of errors. Questions and misunderstandings should be cleared up at once, as an uncorrected error that is repeated is the fault of the supervisor.

In addition to the quality of work, standards of conduct must be high. The supervisor, by his or her attitude and conduct, sets the style for subordinates. A supervisor who is irregular in habit, late for appointments, careless about facts, or bored in attitude will find these qualities reflected in subordinates.

The supervisor must have an initiative to investigate unsatisfactory work and conditions and the courage to take corrective action. Some of the most important facts of leadership are often overlooked or are not understood. People *want* to do good work. Morale is never a problem in an alert, progressive, efficient unit; high morale is never found in a slipshod, substandard, easygoing unit. High standards alone will not produce high morale, but high morale is impossible without them. Ralph Waldo Emerson said, "Our chief want in life is somebody who will make us do what we can."

Acknowledge Good Work with Praise

In Chapter 6 it was shown that praise stimulates good work and may be used as a motivating force. Praise often softens necessary criticism. Praise stimulates by appealing to the basic human desire to feel important. There is a greater danger in too little praise than in too much. It is the rare supervisor who goes to the excess—slops over—in telling people how well they are doing. On the other hand, too much praise or praise given uncritically loses its value; as in all things, care and good judgment must be exercised.

Praise given publicly has extra weight; it raises morale, standing, and self-confidence. The person being praised has additional assurance that the praise is genuine—he or she has witnesses. Other people see that good work is observed and that deserving individuals are not overlooked. Public praise may be either verbal or written. If it is a written commendation, it should be posted on a bulletin board or distributed. The news media may be used where appropriate. The facts for a good story should be made available and the individual's name should be used.

Praise need not be complex or difficult. There are frequent opportunities to convey praise casually. Few supervisors realize how men and women welcome a few honest words of admiration for their efforts. For example: That was a good job of investigation you did on the Jones-Smith incident—Your motorcycle always looks beautiful! How do you do it?—That report was outstanding! I can't think of a single point or lead that you overlooked—That driver was really a hothead, but you did a great job of handling the situation.

Praise is of great importance to the supervisor who wants to be respected. It is evidence that efforts are noted and appreciated. Nobody can keep a chip on the shoulder if the supervisor consistently acknowledges positive performance.

Be Consistent

Consistency is necessary to maintain the momentum of an organization, and most units will operate for a long time on sheer momentum. But when supervisors institute abrupt changes, issue orders that conflict with established policies, or create other inconsistencies, they create confusion. The operation will deteriorate much more quickly as a result of inconsistency than it would without any supervision at all. Orders should be changed only when necessary and then only when the change is well thought out and carefully planned. Overruling subordinate supervisors is extremely poor practice. It should be done only when absolutely necessary and never without an explanation.

Consistency in attitude and manner is also extremely important. A supervisor who is moody—alternately optimistic and pessimistic—is not a leader. Neither is one who is affable and easygoing one day but unfriendly and overly strict the next. Subordinates will not respect such a leader. They will be cautious and, not knowing what attitude to expect, will be bewildered.

Be Patient and Calm—Practice Self-Control

The successful management of people requires self-control. Subordinates will not always understand orders the first time, so they may ask questions to which the answers appear obvious. To a busy supervisor, the time spent clarifying instructions and listening to poorly organized arguments may be irritating to the point of exasperation. But the leader cannot afford to let emotions dictate his or her conduct. Calmness and patience are essential to good leadership. Intolerance and impatience produce undue haste and confusion. These qualities show the supervisor's ignorance of subordinates' ignorance—the lack of knowledge of the extent of misunderstanding possible among personnel.

Undue haste should be avoided. A supervisor should never set a pace that subordinates cannot, or will not, maintain. Caution is also advised when assigning blame; a slower, more analytical approach may prevent a bad mistake. There is a natural tendency to want results immediately. Again, patience pays dividends. It takes a certain length of time for even the most reasonable goals to be met, and impatient harassment of subordinates will further delay—not expedite—the achievement of those goals. Calmness is contagious; the supervisor's attitude and manner will be copied by subordinates. A calm manner is particularly important in emergencies. When men and women watch their supervisor closely, if the supervisor is calm, they will be too.

There is a human tendency toward anger. The body is geared to jungle warfare, with wonderful mechanisms that help it to put up a good fight. When anger is aroused there is an immediate response—the body is prepared to fight, but the brain is not necessarily prepared to think. Furthermore, fighting is usually impractical because our civilization does not condone such primitive behavior. When angry, we must either "keep the lid on" and fume quietly, or "blow our tops," in which case we may say or do things we would never do normally. These things done in anger may never be forgotten, and permanent harm may result. There is no evidence that a person is born with a quick temper. The control of temper is learned. Some individuals are actually proud of having a quick temper. What they are really proud of is the learned ability to stop thinking with their brains and to start thinking with their glands. The latter, of course, is *not* thinking at all; it is reacting. *There is no such thing as a good supervisor with a quick temper.*

For the supervisor, learning to control a quick temper is vitally important. The first step is for the supervisor to impose self-control; because control is learned, each success will make it easier the next time. The goal is never to lose one's temper, regardless of the provocation. In circumstances when loss of temper seems likely, politely but firmly refuse to continue the matter at hand if it is possible to do so. When the supervisor is dealing with an employee, this is a must! Defer the matter, even if only for a short time, until control is assured. It is a curious fact that some supervisors know that this is a good tactic but refuse to do it, fearing that they will reveal a lack of self-control. Yet, how much more prestige is lost in the ensuing flare-up than would have been lost by postponing the argument!

Give Credit for Ideas

Many supervisors fail to credit subordinates for their ideas. They add to or improve an idea before passing it on and believe this gives them the right to take the credit. This is ethically and morally wrong. When any cash reward results (as from a merit award system), it is also legally wrong. Furthermore, any respect gained from superiors by this practice is far outweighed by the loss of respect from subordinates.

Credit for ideas should be given in public if practical—the individual, the individual's associates, and his or her supervisors should be told if the idea is sufficiently important. If the idea is eligible for a merit award, the originator should be encouraged to write it up and submit it. Unselfishness in this respect builds good relationships. It produces trust and loyalty; the subordinates become convinced that the supervisor is fair, honest, and appreciative.

Take Personal Responsibility for Errors

One of the basic principles of organization is that responsibility be accompanied by sufficient authority to perform the job for which the person is responsible. This works two ways: a supervisor who is responsible for a job has the authority to get the job done but is also responsible for the way authority has been used. The supervisor is responsible for any errors which may result from poorly delivered instructions or for errors which may be made by subordinates in interpreting those orders. Some supervisors have an excessive fear of being wrong. They seem to believe that a mistake disqualifies them as supervisors, and they fear that they will lose the respect of both subordinates and superiors. In order to avoid this liability, they try to blame others. Remarks from this type of individual can range from "I told him not to do that!" and "I didn't know that could happen!" to "I can't understand why she would order a thing like that!" and "I didn't tell him to go that far."

Taking the blame for errors or wrong decisions is essential to winning respect. No one expects a supervisor to be perfect. He or she will not lose face if decisions are not wrong too often. Even if the mistake was only partly the supervisor's fault, he or she should take all the blame rather than try to pass it to a subordinate. Oddly enough, this practice increases the respect of subordinates, and the error in judgment is overshadowed by admiration for the supervisor's courage and honesty. In addition, the supervisor must be able to show by voice and manner that he or she really does accept responsibility. A leader must have the ability to maintain poise under stress and criticism.

Be Competent Professionally

To lead, the supervisor must know that no one can bluff all of the time. There is no substitute for exact knowledge. A supervisor cannot hope that competent subordinates will carry him or her indefinitely. Good, prompt answers must be given to questions from subordinates. Such questions usually concern the method of doing various jobs and the correct procedures to be used. The supervisor must have a good working knowledge of all subordinates' jobs. This requires keeping up to date on changes that occur.

Be Businesslike in All Dealings

A businesslike attitude conveys assurance that responsibilities are not being taken lightly. Conducting oneself with dignity does not necessarily mean being solemn. It is possible to be cheerful, pleasant, and friendly, and still businesslike. On the other hand, levity or horseplay, particularly in inappropriate times, will lose respect. Also, profanity, coarseness, and rudeness in speech are not only undignified, but also are not consistent with good leadership. Subordinates may be amused by a clown, but will not respect or accept leadership from such a person.

Make Prompt Decisions

The supervisor who defers decisions is not a leader. Delay is frequently disadvantageous to a project, but the effect of procrastination and indecision on subordinates is even worse; they become confused, not knowing what to do. The clear inference of such delay is that the supervisor lacks either ability, knowledge, or courage. Making decisions promptly does not mean making them hastily; some delay may be necessary to get the facts. Sometimes the supervisor may not know the answer to a question, but if immediate action is taken to get the facts in order to arrive at the answer, subordinates will respect the person. A leader cannot allow a faulty procedure or operation to continue because he or she does not have time to work out something better. If too busy, the supervisor should assign someone else to do it. Ignoring a problem and hoping it will go away is not leadership.

Exhibit Personal Integrity

Integrity is an observance of principles of conduct that are never subordinated to expediency. Any leader must be technically proficient in his or her specialty and leadership techniques. But technical proficiency alone is not enough. Before a man or woman can lead others, the person must learn self-control. The individual must be mature and have moral principles and ethical standards as well as a sense of purpose in life. The leader must actually do what is expected of others, and must not be guilty of obvious wrongs such as the misuse of the department's supplies or equipment. The leader must exhibit qualitites such as promptness, dependability, and courtesy. The leader must demonstrate exemplary personal habits such as neatness of dress and appearance and good off-duty conduct. Employees will not respect a supervisor whose conduct and actions are not admirable, or one who advocates different standards for subordinates than are personally practiced by the supervisor.

Maintain Good Personal Appearance and Physical Condition

Appearance, manner, habits, speech, and social status do not make a leader, but a noticeable weakness in any of these traits will detract markedly from one. The importance of proper dress is often overlooked. An officer should be neat, clean, and well dressed for all business or social occasions. The fact that most duty time is spent in uniform does not mean that a proper wardrobe of civilian clothes is unnecessary. It is not unreasonable to expect proper dress. Government employees, particularly those working in offices, are almost always well dressed.

Physical condition is important for several reasons. Men and women find it difficult to respect a supervisor who does not keep in good physical condition. They know that their job requires strength, speed, and endurance. Most officers are in reasonably good shape, and they resent a supervisor who could not perform physically if the occasion demanded.

Police officers should project a rather glamorous image. The public believes that they are persons of the highest calibre who are capable of skillful violence when necessary. The public looks with disfavor on any member of the department who is obviously not in good physical condition. A supervisor who is not in good condition cannot reasonably expect subordinates to maintain themselves in good condition. A poor personal example from the supervisor who is instructing subordinates to stay physically fit creates an impossible situation. The supervisor who expects to be a leader must exert the self-discipline necessary to control personal weight. Excessive weight comes from just one thing—overeating!

How to Win Confidence

The leader must have the confidence of subordinates. Men and women will not be loyal to a supervisor they don't trust. Lack of confidence in a leader implies a lack of confidence in instructions. Obviously the leader must take advantage of all opportunities to build confidence and must guard against actions that would destroy confidence. The following suggestions on winning confidence will be helpful:

1. Insist on honest, honorable, and proper methods and practices.
2. Face the facts.
3. Keep subordinates informed.
4. Avoid criticizing superiors.
5. Keep promises.
6. Support the valid interests of subordinates.
7. Get things done for subordinates on time.
8. Now and then, help subordinates do their jobs.
9. Listen to subordinates' complaints.
10. Appear confident.
11. Respect the confidence of subordinates.

Insist on Honesty

Reports and testimony must be honest. It is far better to lose a criminal case than to falsify a report or give false testimony. Supervisors must comply with orders and policies and must not attempt to circumvent them by looking for loopholes. Shady and questionable practices and not-quite-true statements destroy the confidence that the leader must have.

Be Candid

A leader who is candid gains the confidence of subordinates. They know what is wanted of them, they know where they stand, and they are not afraid to bring problems or difficulties to the leader for advice or assistance. A supervisor who is frank admits errors openly, meets important issues squarely, and makes quick but thoughtful decisions.

Keep Subordinates Informed

It is essential that subordinates be told as far in advance as possible of any changes that will affect them. Adequate information stifles rumors, makes the men and women feel they are part of the team, and reduces objections to the change. News of changes or proposed plans should be given to line supervisors first, so that they can explain them intelligently to their subordinates and to the public. This practice prevents humiliation and the loss of respect which occur when subordinates know more about operations than do their supervisors. If the supervisor cannot answer a question from a subordinate, the person should not be sent to a higher supervisor to find the answer. The supervisor should go personally with the individual.

Police officers must obviously know more about the operations of their agency than the general public. They will be asked questions regarding newspaper articles, radio and television programs, announcements, etc. If they are not familiar with the matter, they will be embarrassed and will resent the fact that they are not considered important by the department. When employees are informed, efficiency is increased in the following ways: they will think about problems and possibly suggest some good ideas which can be used; and they will avoid activities, statements, or operations that might conflict with current programs.

Avoid Criticism of Superiors

Whether it is justified or not, criticism of superiors destroys the confidence of subordinates in the entire command structure, including the supervisor who is doing the criticizing. The supervisor should seek a constructive explanation for mistakes, oversights, delays, etc.

Keep Promises

The best way to avoid breaking promises is to avoid making them. New supervisors in particular are prone to make rash, impulsive promises which cannot be kept. Some older supervisors never seem to learn the harm which can be done by making promises which are not kept. This may seem to be a small matter, but some authorities in the personnel field believe that broken promises destroy morale more quickly than any other factor. Everyone has had experiences in which promises made to him or her were not kept. The memory and bitterness can last for years, and confidence in the person who broke the promise may be completely destroyed or badly damaged. Subordinates often misinterpret a statement and believe that a promise has been made. "I'll see what I can do" becomes "I'll do it for you if I possibly can," or "I'll do it." This misunderstanding can be avoided by a clear definition of what is meant and what will be attempted. It is better to disappoint the subordinate immediately than to compound the disappointment later and diminish confidence in the supervisor.

If a promise was made in good faith, it must be kept if humanly possible, even if it occasions more work and difficulty for the supervisor than was anticipated. If it is impossible for the supervisor to keep a promise, the reasons should be explained to the subordinate at the first opportunity. If the promise was made in the presence of other employees, the supervisor must also see that they know why it was not kept.

Support the Valid Interests of Subordinates

The leader must show that he or she has subordinates' interests at heart. It must be demonstrated that subordinates will be aggressively supported when they are right and when they *mean* right, but that they will not be supported in derelictions of duty, deliberate violations of the department's regulations, or willful misconduct.

Get Things Done for Subordinates on Time

To win confidence, the leader must do the job properly and promptly. If an undesirable situation develops and is discovered, it must be corrected immediately. If changes or improvements are promised, they should be fulfilled as soon as possible. Promptness and decisiveness of action increase both respect and confidence.

Help Subordinates to Do Their Jobs

When a job is difficult and a man or woman is obviously having trouble, the supervisor should help the person with it. If it appropriate, the supervisor can tell the subordinate that the job is unusually difficult, that the job may require two heads instead of one, or that the subordinate cannot be expected to do the job because he or she has not been properly instructed. The supervisor should not do the job personally. When a competent job of coaching is done, the subordinate not only gains self-confidence, but also has increased respect for the leader's job knowledge. The tact used and the time devoted to helping the employee show that the leader considers him or her important.

Listen to Subordinates' Complaints

The supervisor who has a real interest in subordinates will readily receive complaints. If complaints and grievances are received reluctantly or are deprecated, the individual must feel either that the supervisor is not really interested in subordinates, or that the supervisor is trying to avoid trouble, fearing that he or she may have to back up actions and policies or correct a difficult situation. Welcoming discussion of complaints and grievances helps to detect dissension before it becomes serious and increases the subordinates' confidence in the supervisor.

Appear Confident

A leader must be something of an actor. Everyone has periods of depression when he or she is tired, worried, or discouraged, but the leader must learn to conceal personal feelings and show a confident and assured face to subordinates. The leader must have a strong belief in his or her own competence, as well as the ability to display self-confidence in a natural and acceptable manner.

Respect the Confidence of Subordinates

The subordinate looks to superiors for direction of activities, for authority to act, and for approval or disapproval of activities. An employee expects to be told what his or her job is, what the job entails, and how well he or she is performing. This aspect of the subordinate-superior relationship produces considerable anxiety in subordinates. Employees focus a good deal of attention on the boss—anticipating wishes, being very sensitive to moods, and often finding hidden meanings in gestures that were intended to be casual. This sensitivity occasionally produces overreaction to commands and misinterpretation of requests. What the supervisor says has a special impact on subordinates. An unintended inflection of the voice, a careless choice of words, or by-passing of a subject can cause insecurity that interferes with efficient work. Thoughtless remarks, forgotten in a flash by those who make them, cause many restless nights and frustrating days for those affected.

Unfortunately, most people are more concerned with what their own supervisors think about them than with how they are regarded by subordinates. Many subordinates watch for favorable times and moods to present a request or to pass on necessary information, and as a result the supervisor never receives these messages. The channel of communication that results from this practice works by fits and starts, or not at all. The supervisor is effectively insulated and has only a partial and self-interpreted picture of what is actually going on. Respecting the confidence of subordinates will help solve this problem. The supervisor should welcome the offering of any information or problem. If the matter is confidential, and revealing the source would embarrass the individual, the informant should be protected. If the supervisor avoids moodiness and fluctuations in attitudes and insists that all pertinent information be given promptly, the supervisor lets subordinates know that their confidence will be respected and appreciated.

How to Win Loyalty

Loyalty is essentially an emotional feeling rather than an intellectual commitment. It is not one-sided; subordinates will be loyal to their supervisor only insofar as their supervisor is loyal to them. Friendship plays a large part in the establishment of loyalty, making it hard to imagine persons being loyal to a cold, unfriendly supervisor. The supervisor who refuses to develop friendly relations with subordinates on the basis of some old platitude such as, "Familiarity breeds contempt," will find that in emergencies, when fast, blind acceptance of instructions is necessary and loyalty indispensable, the loyalty required simply isn't there. Friendliness does not imply weakness. Nor does it imply a lack of discipline. It does imply an interest in and a caring for the individual man or woman. Taking all these factors into account, the supervisor who wishes to have the loyalty of subordinates will be friendly—but firm—as well as loyal to and interested in each of them.

Be Pleasant

Cheerfulness is friendliness, conveying a liking for others. Grouchiness implies and conveys dislike. Cheerfulness stimulates faith in oneself, the department, and the job. It produces a pleasant environment in which more work can and will be done. It helps overcome obstacles, produces better cooperation, and smooths relations.

Cheerfulness and friendliness can be consciously developed. Like everything that is worthwhile, this requires some effort and self-discipline. It is more difficult for those who have not developed these qualities as part of their personalities. Developing cheerfulness starts with a simple muscular operation—a smile. This requires acting if a person is not really feeling cheerful, but it more than pays off in smooth relations. Cheerfulness and friendliness are not undignified. It is the uncertain, apprehensive, and insecure supervisor who is afraid to be cheerful and pleasant.

Be Available

A supervisor who is distant and difficult to talk to does not encourage friendly relations with subordinates. The supervisor must show that he or she welcomes consultation on any problem or concern and must develop an easy conversational relationship with subordinates. This is important for three reasons: (1) new ideas can be offered casually, even if the subordinate is not sure of their value, and even if all the details are not worked out; (2) the channels of information, so necessary to a supervisor, will be open; and (3) the supervisor-subordinate relationship becomes easy, informal, and workable. The third reason is the most important one. There are several techniques and practices which will help the supervisor seem available. A cheerful greeting and a few easy words show that the supervisor is available in both body and spirit. Many personal contacts characterize the effective superior. When employees want to talk about a problem at the office, the supervisor should consider not talking across a desk, because it is a symbol of authority and may create a barrier to free discussion. By removing the barrier of the desk, the supervisor makes effective communication possible.

Be Sympathetic

Many modern personnel administrations are placing great emphasis on "empathy" as being absolutely indispensable to a leader. Empathy is the imaginative projection of one's consciousness into another human being. In simple terms, it is the ability to see a problem from the viewpoint of the other person or to put oneself in the other person's shoes. The leader must be sympathetic with a subordinate's troubles. When people have personal or business problems they crave sympathy and, by simple transference, like those who sympathize with them. Conversely, they resent those who fail to sympathize. Therefore, supervisors must never become so preoccupied with duty that they fail to respond to the troubles of subordinates. *Any problem that is important to a subordinate must be important to the supervisor.*

Recognize Subordinates as Individuals

Employees are people as well as workers. The department hires the whole person, not a fragment. This includes the person's hopes and plans, financial and domestic problems, and personal history—training and experience, likes and dislikes, and special abilities and deficiencies. It is possible to direct the activities of subordinates while knowing little about them, but this is not leadership. Developing cooperation and job enthusiasm in a man or woman requires appealing to the person's beliefs and interests, and can be achieved only by taking a genuine interest in the individual, by talking with and sincerely getting to know the person.

A real interest should be shown in each person. A leader who is really interested in subordinates will find that they are interested in their leader. Ambitious employees always make a point to be informed about the things their superiors are interested in so they will be liked and considered for advancement. The same thing works in reverse. Subordinates will like and work harder for a leader who is interested in them as individuals.

To produce loyalty there must be affection; to produce affection there must be interest; to produce interest, there must first be interest in the subordinate by the supervisor. The reasons for showing a real interest in the ideas and suggestions of subordinates should be obvious: their thoughts may be valuable; interest shown by the supervisor will stimulate future thinking by subordinates; and it is an easy way to further friendly relations. Interest and friendliness must be sincere. Insincerity will be easily detected and will only produce antagonism. The supervisor can demonstrate interest in subordinates by simple measures such as using first names, knowing and asking about employees' families, holding short, informal "bull sessions," and having lunch with subordinates on occasion.

Be Concerned for the Health of Your Subordinates

Concern for health and safety is a supervisor's duty. It prevents economic loss to the department through lost time, the training of new employees, and the destruction or damage of supplies and equipment. Such concern gives the supervisor a splendid opportunity to develop loyalty, if the supervisor is positive but tactful in the methods utilized and shows that he or she is genuinely concerned with the well-being of personnel. The supervisor must not allow an employee to work when ill. It is not good for the individual, and it is not good for others working with this person. For example: the heroic employee who comes to work with a bad cold is prolonging the illness, increasing its severity, and infecting coworkers. Encourage and assist employees to get adequate medical care. Take the steps necessary to insure healthy working conditions in their offices, such as fresh air, proper temperature, cleanliness, good lighting, and freedom from unnecessary noise. *Anything that affects the employee's efficiency during working hours is the supervisor's business.* This is of particular importance for uniformed personnel, because dulled reflexes or vision can cause disaster. Watch for the following: excessive drinking, insufficient sleep, and excessive fatigue from outside employment.

In making work assignments, health factors must be considered by distributing the workload as evenly as possible and scheduling work to avoid rush jobs. When tension is added to a heavy workload, fatigue increases. When possible, assign people to jobs they like to do; they will get less tired.

Employees should relax whenever possible. The supervisor should see that they take their lunch hours (and coffee breaks, if assigned to office duty). Studies have proven the value of rest periods in maintaining health and improving efficiency. When possible, give vacations at the best times for the employees.

Maintain Standards of Safety

Accidents do not merely happen. They are caused, and many of them are preventable. Until each police supervisor and each individual officer look upon working safely as an individual responsibility, the police supervisor will continue to be plagued by preventable accidents. The good supervisor can make safety in all phases of police operations an important departmental objective, so that the entire work group will regard safety as both a group and an individual responsibility.

The supervisor sets the pattern for the practice of safety. For example, if a sergeant is known to disregard the use of seat belts in patrol cars, he or she is in no position to criticize a subordinate who is injured in an automobile accident because the seat belt was not fastened. The operation of motor vehicles by police personnel is a major concern of management and probably the greatest single hazard to safety in the police service. Laws and the department's rules and regulations usually govern the operation of emergency vehicles, and it is the responsibility of the supervisor to see that subordinates adhere to these rules. The officer who is allowed to "hot rod" habitually is going to be involved in an accident sooner or later. Instruction in defensive driving is a *must* for officers operating vehicles, and supervisors should see to it that their men and women practice what they learn in training courses.

The nature of police work is such that, of necessity, it involves more hazards than many other types of employment. However, the supervisor who insists that subordinates adhere to operational techniques designed to minimize the hazards will have a lower accident rate than the supervisor who is lax in enforcing safety precautions. It is not practical to discuss all of the areas of danger to police officers, but a few examples are firearms, apprehension and transportation of criminals, effecting various types of rescues, and traffic control. It is not possible to eliminate all accidents in the police service, but it is possible to hold them to the minimum number that would occur no matter what precautions were taken.

Responsibility for Accidents

Assuming that the supervisor has properly instructed subordinates in safe working methods and has enforced safety rules, the individual officer is primarily responsible for most accidents but from the viewpoint of management, the supervisor is responsible for the accidents in his or her unit. The supervisor represents management in teaching safe operating practices and in seeing to it that subordinates follow these practices.

Management expects the supervisor to enforce safety rules as strictly as other types of standards are maintained. The supervisor is supposed to teach subordinates how to work safely, lead them in the practice of safety, and exact penalties for a failure to work as instructed. Most of this leadership is of a positive nature, but there are occasions when penalties are called for. The supervisor should make a complete investigation of all accidents within the unit to determine causes, and should take proper action to prevent recurrences.

Safety and the New Officer

The new officer in the department probably has not had any prior police experience. He or she is faced with the strain of adjusting to new associates and a new environment as well as learning a new job. To this individual, nearly everything is going to be new, and it is the special responsibility of this person's supervisor to see that the individual is given detailed, careful instruction in safety. Even though the new employee has been through a recruit school, nothing should be taken for granted, especially safety instruction. Industrial studies indicate that accident rates are higher among new employees than among skilled, older workers in spite of the new workers' eagerness to learn and willingness to observe safety regulations. Most of these accidents occur because of lack of skill and improper or inadequate training.

The attitudes of the young officer toward working safely are acquired primarily from supervisors and fellow officers. If safety is deeply embedded in the consciousness of the work-group, the new employee soon acquires the same attitude. The reverse is also true; if the old-timers give a recruit a bad time about the use of proper procedures, the new officer is likely to ignore the safety practices and may eventually have an accident.

Be Courteous, Tactful, and Considerate

Courtesy is polite behavior and thoughtfulness of others. It is a great defensive tool because it begets courtesy, increases self-control, and produces respect. Courtesy is not deference; employees in the highest positions are courteous to subordinates. A supervisor must be as courteous to subordinates as he or she is to superiors and strangers. Subordinates should not be interrupted when talking any more than strangers should be. The same expressions of courtesy, such as "Please," "Thank You," "May I," and "Could You," used with strangers should be used with subordinates.

The failure to remember names is a discourtesy. Although this becomes increasingly difficult as higher rank is attained and an officer has more subordinates, it also becomes increasingly important. One whose name is remembered feels important; one whose name is forgotten feels unimportant, discouraged, and sometimes resentful.

Tact is simply regard for another's feelings. It is the selection of approaches, words, and phrases that will promote positive moods and feelings, such as loyalty, justice, fairness, courtesy, and affection. Tact is also the avoidance of approaches, words, and phrases that provoke negative moods such as hatred, suspicion, fear, resentment, and anger. Tact is *not* the avoidance of unpleasant subjects, it is not sidestepping disagreeable, distasteful subjects that must be faced.

Be Impartial

Prejudice is a conclusion or opinion based on insufficient evidence. Supervisors must not form an adverse opinion of a subordinate without good cause, and must not let a single action of a subordinate create a prejudice. If a decision affecting an individual subordinate has to be made and the supervisor believes he or she is prejudiced against this subordinate, the supervisor should discuss the problem with an impartial individual whose judgment is respected.

Favoritism is the giving of a greater number of privileges to an individual than the individual's performance or position warrants. Supervisors are accused of favoritism more often than they deserve because they often fail to justify their choice for desirable assignments when they are made. Even though the supervisor has the authority to give assignments without explanation, the arbitrary use of this authority is damaging to the morale of subordinates. An obvious solution to this problem is to discuss the reasons for various choices at convenient and appropriate times. Meetings and conferences in which the subordinates are present offer excellent opportunities for this approach. It is sometimes possible to include the reasons for an assignment in the written order and have it distributed to all the employees or to discuss the issue with key subordinates—the employees' natural leaders—who will carry it back to them.

In situations that require disciplinary action, the supervisor must be especially impartial. The supervisor must get all the facts and be willing to look for faulty instruction, poorly conceived procedure, and inadequate supervision. He or she must consider the welfare of both the individual concerned and the police agency.

Friendships

The supervisor must handle friendships with subordinates carefully. Many supervisors are wary of social contacts with subordinates, because of quarrels that have developed at card games, parties, or sporting events, or because of subordinates who have tried to take advantage of such contacts. The supervisor who is too familiar with subordinates loses a very valuable supervisory advantage. It is difficult to snap back to a business relationship after a period when all barriers are down. As previously stated, the supervisor must guard against favoritism. Considerable social contact with one or two members of a group will be regarded as favoritism and resented. It should be possible to develop the right sort of friendliness during business hours while maintaining the necessary barrier.

Rule: Be one WITH, but not one OF, the men and women.

Certain exceptions must be recognized. Whenever there is a social gathering of or for the entire group, the supervisor must attend. He or she may leave early or when good judgment indicates, but a failure to appear will hurt the supervisor. The usual reaction of subordinates to the supervisor who does not attend such social events is that this person is not friendly and does not care for them, a feeling that will be reciprocated.

There are no clear-cut rules about friendships with subordinates at a level close to that of the supervisor. The best policy is to go as far as seems safe and to stop short of the point at which subordinates threaten to take advantage of the relationship. The supervisor who is very friendly with a subordinate several levels below his or her own must confine the relations to off-duty times. Otherwise, intermediate supervisors are placed in an awkward position. The most effective technique for maintaining the necessary barrier is to insist on being addressed by title in all official relations, when on duty or when business is being discussed. The practice has the psychological effect of serving as a subtle reminder of the proper relationship.

DISCUSSION QUESTIONS

1. How is leadership established?
2. Why is it vital to have the respect of your subordinates?
3. Why is it a sound practice to give praise publicly?
4. In what ways is consistency important in supervision? Why?
5. How does the supervisor go about retaining control of his or her temper? What is wrong with blowing off a little steam now and then?
6. Who should be held responsible for errors? Why? Can you "pass the buck" for mistakes?
7. What are the qualities of a good supervisor? Why is each important?
8. How should the supervisor go about winning the confidence of subordinates? Why is this important?
9. Why is it important to keep subordinates informed? What happens when a supervisor fails to keep subordinates informed?

LEADERSHIP

10. Under what, if any, circumstances should a supervisor support subordinates? Should the supervisor support management against subordinates, right or wrong? If not, what should the supervisor do about it?
11. What is so important about an attitude of confidence?
12. How does the supervisor gain the loyalty of subordinates? Why is this important?
13. When would the personal, off-the-job habits of a man or woman become a matter of concern to the person's supervisor? Why? What action should the supervisor take, if any, to correct off-the-job problems?
14. What is the supervisor's responsibility for safety? Why is this so important to the department? How can the supervisor prevent accidents?
15. What should the supervisor's attitude be toward social friendships with subordinates? Why?

8 Morale

The personal adjustment and morale of employees is one of the most important considerations of the modern executive. In the police service, morale is even more important than in comparable business enterprises because poor morale in the police service not only affects the individual and the department, but the general public as well. The police supervisor must, therefore, be concerned with the establishment of high morale and with the personal adjustment and happiness of employees. Should the supervisor feel inadequate in this area, he or she must learn very quickly to deal with this problem, as the employees themselves and the associations which represent them are very conscious of the fact that poor employment conditions need not be tolerated.

In order to maintain morale at a high level, the supervisor must first know what it is, and must understand its importance and the extent of his or her responsibility for its maintenance. Secondly, to raise morale, it is necessary to understand the needs and wants of people, the methods of satisfying them, and what constitutes personal adjustment. Finally, since there is a definite relation between morale and efficiency, this relation must be explored, along with the methods of determining the actual morale level.

What is Morale?

Morale is an intangible quality similar to and perhaps embodying courage, zeal, and confidence. The French call it *esprit de corps*. It is to the mind what condition is to the body: fitness for the task at hand.

Some people feel that morale should be distinguished from job satisfaction. They feel that morale is a spirit of devotion to the endeavor of a group that grows from mutual feelings that the goals of the organization are good and that its members' contributions are appreciated. Thus, when the morale of the group is high, a zest for living may be shared even by an individual whose job satisfaction is low. The sense of belonging in such an organization compensates for the lack of job satisfaction. The group holds together because of its internal bonds, and is independent of outside pressure. The members are enthusiastic about the leadership, goals, and perpetuation of the group.

Others consider morale to be an elusive quality for which there are as many definitions as there are those willing to define it. Morale obviously means different things to different people.

Because of the difficulty in defining morale, some authorities prefer to adopt an operational definition: high morale is a complex combination of many factors that make people want to do what the organization expects them to do. Conversely, low morale is a combination of factors that prevent or deter people from doing what the organization expects them to do.

The Importance of Morale

It should be obvious that merely teaching a man or woman the skills necessary to be a good police officer and providing the person with a uniform and equipment are not enough. Qualities such as enthusiasm, personal satisfaction, and a willingness to work as part of a team—the components of morale—are also essential for the continued success of the department's operations. Only if each officer in the department is interested in and satisifed with his or her job will there be the wholehearted and cheerful cooperation that is essential to the functioning of the organization.

Morale and Productivity

Morale and productivity are related. When morale is high in a department, there is likely to be less absenteeism and turnover than when morale is low. People who are interested in their jobs and find their working relations with supervisors and fellow officers pleasant are less likely to be absent from work or to quit than are individuals who work under less favorable circumstances. High morale, therefore, means that at any given time there will be more personnel available for police duties and lower costs to the department in terms of training budgets.

Many psychological studies have indicated that one of the leading factors in high employee turnover is poor morale. When we consider the cost of training a new police officer, it becomes apparent immediately that low morale is much too expensive for a public agency.

Public Relations and Morale

Public relations and morale are related; the attitude of officers toward the department will soon become known in the community. As police officers associate with friends, neighbors, and the public at large they cannot help revealing their attitudes, both favorable and unfavorable. Criticism expressed by a small minority of officers can spread through the community and offset the best efforts of the majority to build good public relations for the department. Well-qualified men and women will not apply to a department when they feel that they would be unhappy in it.

Factors in Morale

Morale is not something that can be dictated by management or built overnight. It is developed over a long period of time as a result of sound personnel policies and procedures, good supervisory practices, and other factors. Influences on morale should receive careful attention from police management, and a continuous, positive effort should be made to build good morale. While it is important to correct undesirable conditions, it is more important to place emphasis on positive action to develop the type of work environment that will contribute to high morale.

Morale is closely associated with discipline. Since the attitudes and feelings of each member of the group influence the behavior of other group members; the supervisor must be concerned with the conduct and discipline of subordinates. This is done by establishing and publishing reasonable standards of conduct and by enforcing them wisely. Such practices are conducive to good morale. The group will enforce the standards itself by applying social pressure to members who get out of line.

It is a basic management responsibility to have an established procedure for handling employees who fail to conform to the standards of performance and conduct that are expected of them. There also must be a procedure by which employees can express their complaints and grievances to management with the assurance that they will be given careful consideration.

Morale and Police Efficiency

The importance of morale in police efficiency is obvious. Officers with high morale tend to approach their jobs and the public with a more positive viewpoint, and thus encounter fewer conflicts with the public. They have more desire to succeed in the accomplishment of their tasks and respond more willingly in stress situations. They will remain with the police service longer, thus giving the organization and the community the benefit of their years of training and experience. High morale is certainly as important as good training and equipment.

Supervisor's Responsibilities in Maintaining Morale

From the time a man or woman is recruited into the department until the day the individual leaves, a large part of his or her life is influenced by the personnel policies and procedures of the administration, by the supervisor he or she works under, and by relations with fellow officers. It has been said that being a good police officer is not just a job, but a way of life. All these influences, together with all an individual's life experiences, will determine how well the person will adjust to the job and to daily life.

By establishing policies and procedures that facilitate an employee's adjustment, the department not only contributes to the mental health of the officers but also improves efficiency and public relations, and reduces turnover. The good supervisor will be able to recognize the symptoms of emotional problems in subordinates and will take steps either to handle the problems personally if they are minor, or to refer them to professionals if they are serious.

The supervisor is responsible for developing and maintaining good working relations with subordinates as well as the highest possible level of efficiency among subordinates. The effectiveness of the department depends largely on how well supervisors measure up to such responsibilities.

One of the supervisor's principal values to the department is an ability to create and maintain the working relationships that will motivate subordinates—both as individuals and as members of the group—to put forth their best effort. Each employee represents a large financial investment to the department. To protect this investment, the department, through its supervisors, must make every effort to help each individual become and remain an efficient, productive, and satisfied worker.

To these ends, the supervisor has the responsibility of seeing that each subordinate knows the following:

1. The objectives of the job and of the unit in which the individual works

2. The tasks to be performed

3. The accepted methods of performing these tasks

4. The standards of job performance

5. How well the individual meets those standards of performance

6. How the individual can improve his or her work and develop further capabilities

7. The policies and regulations that govern the officer's work

8. What is considered proper conduct or good discipline in the work group

Human Wants and Needs

One of the first steps in raising the morale of an organization consists of finding out what the members of that organization want and need. If wants and needs go unsatisfied, morale is bound to be low. There are some wants and needs that are common to almost everyone; others are highly individual. The individual wants or needs are in a special category, and the supervisor will have to deal with them on an individual, special basis. Such needs might be for particularly expensive medical care for a member of the family, or for an unusual leave of absence.

The common needs, such as the need for food or for recognition, are not always clearly defined, but they are distinct enough to be discussed in a general fashion. To facilitate discussion, the classification of human needs and wants developed by A. H. Maslow has been used.[1]

The Physiological Needs

In this group are the needs for food, water, air, rest, etc., that are required for proper body functioning.

The Safety Needs

In this group are the needs for safety and security in both a physical and a psychological sense. The need for protection from external dangers to the body and the personality are included in this category. Most employees want to work at jobs that provide a reasonable degree of freedom from physical and psychological hazards and a reasonable degree of job security.

The Belonging and Love Needs

The needs for attention and social activity are the major ones in this category. An individual wants affectionate relationships with people and desires a secure place in his or her group.

The Esteem Needs

Included in this group are the needs for self-respect, strength, achievement, adequacy, mastery, and competence that are necessary for confidence, for independence, and for freedom. This category also includes the desire for a good reputation or prestige and for the respect of other people.

The Self-Actualization or Realization Needs

These needs refer to a person's desire for self-fulfillment, the individual's dreams of actually realizing full potential, of achieving all of which he or she is capable.

The priority of needs must be established. Since almost all behavior is multi-motivated, it is not wise to focus attention on any one need to the exclusion of the others. Several needs usually demand satisfaction at the same time. However, needs may differ in relative importance to the individual. Human needs are arranged in a priority sequence above. The physiological needs are the most fundamental, and they must be satisfied first. Once the physiological needs have been satisfied, safety needs become predominant, then the belonging and love needs, and so on down the list. The supervisor must recognize that the need pattern of each individual is different. These differences should be recognized and the knowledge utilized to provide satisfaction for the individual as he or she works toward the accomplishment of the department's goals.

The development of high morale depends upon taking proper action to satisfy as many of the employees' wants and needs as are consistent with the goals of the department. It should be noted that much of a person's motivation is at the subconscious level, and for this reason the person may not be clearly aware of his or her own needs. On the other hand, wants are the conscious desires of the individual for those things or conditions that are felt will give satisfaction.

While the wants of the individual go far beyond his or her job, the nature of these wants, as far as the job is concerned, is of vital interest to management in the development of high morale. Psychologists have done many surveys to determine what satisfaction people want from their jobs. It has been found that the type of job a person holds, the economic and social conditions at the time of the survey, the length of time on the job, and a host of other factors all appear to have some effect on the results of the survey. Because of the differences in survey results, it is not really possible to make any conclusive statements about the priority of employees' wants. However, it has been noted that in all cases when supervisors have been asked to rate what they feel are the wants of their subordinates, the rankings of the supervisors vary, often considerably, from the subordinates' own rankings. The following is a fairly typical set of results in ranking ten job conditions:[2]

	Worker Ranking	Supervisor Ranking
Full appreciation of work done	1	8
Feeling "in" on things	2	10
Sympathetic help on personal problems	3	9
Job security	4	2
Good wages	5	1
Work that keeps a person interested	6	5
Promotion and growth in company	7	3
Personal loyalty to workers	8	6
Good working conditions	9	4
Tactful disciplining	10	7

Table 2.

The difference between the supervisor and the worker points of view is important in that it should indicate to the alert supervisor that he or she must study subordinates and avoid the false assumption that the wants of subordinates are the same as those of the supervisor.

Satisfying Human Needs

Finding ways to satisfy human needs becomes a real challenge. Ordinarily, an external incentive must be used as a satisfier. These are of two types, positive and negative. Normally, positive incentives, such as pay and praise, are used because they bring pleasure and satisfaction to the individual and are effective in achieving the department goals. However, at times it is necessary to utilize negative incentives, such as fear and punishment.

Positive Incentives

Money   In our society, money is the incentive most frequently used to stimulate individuals to produce more. However, when employees are asked to list what they would like to receive from their jobs, pay is seldom at the top of the list.

Security   Most men and women want to feel secure in their jobs. They want to know that they are protected from the loss of their jobs and earnings that could result from accident, illness, arbitrary firing, or other causes. They are concerned also about security following retirement. Thus, security is a positive incentive to work. The reasonable security in progressive police agencies that is provided by good civil service systems stimulates employees and encourages them to work more wholeheartedly. They are able to direct all energies toward the job rather than toward the achievement of security.

Praise and Recognition   It is important to men and women that they be recognized and praised for a job well done. However, it is important that praise be reserved for those instances when it is really deserved and can be given with sincerity. Used too frequently, praise loses its effectiveness; used too infrequently, it causes morale to suffer.

Competition   An employee may compete in many ways: with other employees, as a member of a group competing with other groups, or against his or her own performance record. Competition between individuals or groups may be useful in stimulating safety or increasing work output, etc. The use of progress charts, such as those that show traffic control activities in relation to accident rates, may stimulate competition between squads to lower accident rates within their districts. The development of interest in and enthusiasm for high performance requires good human relations and communication to the employee of the performance level desired. It is not inconsistent with good human relations to establish standards and to promote wholesale competition to meet those standards. A knowledge of the results of competition serves as an incentive to better performance and facilitates the learning of the job.

Participation is recognized as one of the best incentives for stimulating the employee's productivity and providing job satisfaction. In addition to providing opportunities for subordinates to participate in meetings and conferences, to serve on committees, or to offer suggestions, the supervisor needs to give attention to employee participation in decision making about the job itself, and must define the conditions under which this decision making is to be done within the work group. Because employees have a personal stake in the success of a procedural change, their participation in decisions will minimize the resistance offered to changes the supervisor wishes to initiate. When practicable, the wise supervisor will consult subordinates and heed their counsel before effecting any change in operations or procedures.

Negative Incentives

Reprimand and Punishment A reprimand may be either verbal or written and may become part of the employee's permanent personnel record. The imposition of overtime duty without compensation, suspension without pay, loss of some part of weekly or annual leave, demotion to a lower rank or pay classification, and separation from the department are examples of punitive actions that can be utilized by the supervisor. Nonpunitive action includes all efforts, short of actual punishment, made by a superior to correct a weakness in a subordinate. When proper use is made of nonpunitive forms of discipline, it is seldom necessary to apply punitive measures. This subject is treated in depth in the next chapter.

Human Adjustment

Adjustment is a basic necessity of life. People continuously make adjustments to a broad variety of situations in an effort to satisfy their needs and maintain an emotional equilibrium. They are not so much concerned, at the conscious level, with satisfying physiological needs as with the satisfaction of emotional needs—belonging, esteem, and self-realization. It is often maladjustment which aggravates the frustration of these needs, and such frustration in turn increases the problem of adjustment. Because the satisfaction of these needs is dependent to some extent on persons in a position of authority (supervisors), it is important that the supervisor attempt to understand the nature of adjustment and maladjustment.

The Well-Adjusted Person

Mentally healthy persons seem to meet the following criteria: They feel comfortable about themselves and are not overwhelmed by their emotions—fear, anger, love, jealousy, guilt, or worry. They take disappointments in stride and have a tolerant, easygoing attitude toward themselves and others; they can laugh at themselves. These people neither seriously overestimate nor underestimate their ability, and they can accept their own shortcomings. They have self-respect and are able to deal with most situations. Well-adjusted people get satisfaction from the simple, everyday pleasures. They feel right about other people, are able to give love and consider the interests of others, and their personal relationships are satisfying and lasting. They expect to like and trust most people, and expect that others will like and trust them. Their self-confidence allows them to respect the differences among people. They do not push others around, nor do they allow themselves to be pushed around. These people feel that they are part of a group and have a sense of responsibility toward other persons. Well-adjusted persons are able to meet the demands of life and to confront and resolve problems as they arise. They accept responsibility, welcome new experiences and ideas, and shape their environment whenever possible, adjusting to it when necessary. They plan ahead without fear of the future, make use of their natural capacities, and set realistic goals for themselves. They think for themselves and make their own decisions. They put their best effort into what they do and obtain satisfaction from doing it.

Good adjustment does not imply that the individual is free of all life's problems, but it does imply an emotionally mature orientation to life that enables the person to solve personal problems in a constructive manner and to weather storms as they are encountered. No one is free of problems. However, some people are better equipped to handle them than others.

The Maladjusted Person

Problem employees are likely to be employees with problems. Emotional disturbances may play a major role in such problems as chronic absenteeism, accidents, high turnover rates, grievances, alcoholism, dishonesty, and the many different forms of job dissatisfaction commonly found in employment situations.

The crank, the bully, the chronic complainer, and other types of problem employees who often demand a great deal of the supervisor's time are usually people who are having trouble adjusting to the world about them. They are usually problems to themselves as well, and often feel uncomfortable about their own behavior. However, the police supervisor should never fall into the trap of labeling as "maladjusted" anyone who merely happens to hold a viewpoint different from that of the supervisor.

Determination of Morale Level

The determination of morale level is a responsibility of a conscientious supervisor. It is not as easily done as reading a barometer or thermometer. Surveys and questionnaires are sometimes undertaken to measure morale. Some studies are helpful in the discovery of morale problems; others merely serve as a sounding board for gripes, complaints, and grievances and, because of improper follow-up, are a waste of time. A check list is sometimes helpful. The following are indicative of low morale:

1. Excessive and unwarranted use of sick leave
2. High turnover rate
3. General deterioration of appearance:
   a. Neglect of haircuts
   b. Condition of uniforms
   c. Personal cleanliness
4. Increased bickering and arguments
5. Careless treatment of equipment
6. Carelessness in report writing
7. Formation of cliques
8. Circulation of rumors

Normally, high morale results when the conditions below exist. A supervisor should compare his or her personnel against the following points:

1. The purpose or objectives of the unit are well known to all personnel.
2. Individuals are treated fairly in disciplinary actions.
3. The group enjoys a reasonable sense of security.
4. The training within the group is good.
5. Recognition is granted for accomplishments.
6. Rumors are eliminated promptly.
7. Maximum information on problems is provided to all levels.
8. Supervisors display strong leadership.
9. There is complete delegation of authority commensurate with responsibility.

10. Lower-level employees participate in organizational decision making.

The supervisor should remember that high morale cannot be something dictated by management—it cannot be built overnight. Morale must be developed over a long period of time as a result of sound personnel policies and procedures and good supervisory practices.

While it is important to correct undesirable conditions, it is more important to employ positive action to develop the type of work environment that will contribute to high morale. High morale is developed by satisfying as many of the employees' needs and wants as are consistent with the goals of the department.

DISCUSSION QUESTIONS

1. What are the five basic types of human needs? Explain each of them.
2. What is the priority sequence of the different types of needs?
3. What is the importance of this priority sequence in supervision?
4. What are the commonly used incentives for satisfaction of needs? How are they used in police management?
5. What are the criteria that are met by mentally healthy persons?
6. What are some of the problems likely to be caused by emotional disturbance? How will they affect the worker on the job?
7. What is the importance of morale? How does it affect public relations?
8. What are the factors that influence morale? What role does the supervisor play in developing high morale?

NOTES

1. A. H. Maslow, *Motivation and Personality* (New York: Harper & Row, 1954), pp. 91-92.
2. William B. Melnicoe and John P. Peper, *Supervisory Personnel Development* (Sacramento, Calif.: California State Department of Education, 1965), p. 87.

9 Employee Groups and Organizations

In our everyday life, each of us belongs to many different groups. The strength of our ties with each of our group memberships will vary depending upon our own perceptions and the degree of loyalty we feel to the group in question. For example: Most of us identify ourselves as members of a family group, a social group, a work group, and a particular ethnic group. Our behavior varies somewhat, depending upon the group we are with or the role we are playing at a particular time. The behavior of the average police officer working a beat will be different from this person's behavior when he or she is acting in the role of a parent. Most police officers have a very strong identification with their work group. In fact, one would be hard pressed to find a more closely knit group than that of police officers. Among the many reasons for this phenomenon are such things as: the mutual dependence of police officers for personal safety; commonality of interests and, often, personality types; societal attitudes that frequently require officers to turn to fellow officers for the majority of their social contacts; and a mutual interest in performing a service to society.

A group is made up of two or more people who in one way or another interact or transact with each other. If the group consists of more than two people—and most do—it is probable that subgroups will form. Some of these subgroups will remain within the original group, but others will eventually become separate and distinct from the original group.

Within the field of supervision, the concept of the group is of prime importance. The ability of a police agency to carry out its responsibilities and meet its mission and goals is dependent upon the willingness of work groups to follow the procedures and policies prescribed by top management and implemented through first- and second-level supervisors.

There are many different kinds of groups, and the study of group behavior has been the subject of much research and writing. In this chapter, the authors will confine themselves to a discussion of group behavior and its effects on supervision within the police agency.

Typology and Group Characteristics

The *formal work group* is one which has been structured by police management and which has been given a formal charge or responsibility to complete some specific job or task. This may be for a long- or short-term period of time. An example of a formal work group is the day shift in a patrol division, which is charged with beat patrol within an assigned time frame and covering a specific geographic area. Formal work groups will usually exhibit the following characteristics:

1. There is a designated leader.
2. There are specific duties.
3. There are written rules.
4. There are rights and responsibilities.
5. There is an assigned job or task.
6. There are expected standards of performance.
7. There are agency or organizational rewards and punishments.

The *informal work group* is one which coexists with the formal work group but exhibits a somewhat different set of characteristics:

1. There is an informal leader.
2. There is emerging behavior.
3. There are separate norms which may be in conflict with those of the agency.
4. There is reacting and responding behavior.
5. There is a grapevine or "rumor mill."
6. There are peer sanctions.

Some characteristics and processes seem to be common to all groups, formal and informal. These have been listed as follows:[1]

1. All groups have leaders.
2. All groups have followers.

EMPLOYEE GROUPS AND ORGANIZATIONS

3. All groups seek to get some "work" done.

4. All groups have feelings and ideas about how work is to be done.

5. Most groups operate within the minimal requirements of the organization within which they exist.

6. All groups exhibit behavior above and beyond the minimal behavior required to meet the organization's standards.

These six conclusions about group behavior are indicators of the dual character of groups. In formal groups, the structure of the group is dictated from outside the group. Who is to be the boss and how many followers there are, what kind of work is to be done, how it is to be done, the kinds of input expected, and the kinds of rewards and punishments given are all part of what can be classified as an authority system.

The other side of the coin from the formal authority structure is behavior that generates from an informal, unofficial type of organization. There is a designated, recommended, and expected type of behavior coming from the authority source, and there is also the emerging, reacting, or responding behavior in the group that may or may not match the expected behavior. Almost all groups have both of these dimensions working within them, and an individual's behavior in a group is a product of his or her reaction to both of these forces.

It is interesting to note that at least one writer in the field, Elton T. Reeves, identifies a third group type—the semiformal group.[2] He characterizes this type of group by the fact that membership is voluntary, such as in a lodge, church, or other organization. He goes on to state that membership in a work group may be the deciding factor that causes an individual to join a particular lodge, private social club, etc., and that this affiliation with the semiformal group often impinges on the work group by satisfying individual needs and by assisting the person to reach a level of self-fulfillment not possible in the work situation. The effect of such membership can be positive in the sense that it can make a person more tolerant of unpleasant working conditions on the job. But it can also be negative when it satisfies ego needs that should be satisfied at work, thereby leading to a lack of motivation to participate in work activities.

Functions of Groups

Group membership should help in satisfying the individual's expectations and can be said to perform two main functions for individuals: a production function and a social function. When people join work groups, they voluntarily give up many individual rights: they agree to show up for work at a given time, to abide by the rules and regulations of the organization, to work within the structure of the organization, to conduct themselves with some degree of responsibility and honesty, etc. If they violate any of the rules and regulations, they are likely to join the ranks of the unemployed. When individuals join an organization and start on a job, they have to integrate their work with that of other persons. A group exists when all the people within a unit are working for some kind of mutual goal.

What new members in a group want from the group is the following:

> 1. They want the group's help in producing some kind of measureable product, since the organization will reward or punish them in terms of what they produce.
>
> 2. New members want the group's help in achieving satisfaction in what they do.
>
> 3. They have a need to grow as individuals, and expect the group to provide the latitude for growth and development.

The relationship works two ways, however, so the group has certain expectations of the individual and imposes group norms on each member. Some of the group expectations might be stated as follows:

> 1. The group expects individuals to obey its "house rules," which *may* be quite different from those of the larger organization of which the group is a part. Any group has some idea of what should or should not be done by its members and some sort of notion about the range of acceptable behavior for group members. Expectations for newcomers to the group are quite likely to be more rigid than those for longtime group members. Every group has definite house rules or norms, and violation will bring the imposition of negative sanctions or punishment of some sort.
>
> 2. The group expects its members to support the sanctions that have been developed. These may range from physical violence to very subtle social sanctions—for example, freezing out or cooling. Members must accept the value even of extreme sanctions, because they are important to maintaining the character of the group.
>
> 3. The group expects individuals to give up some of their personal liberties and subordinate them to the group, and group members are expected to be "team players" rather than individual "stars."

The Effects of Group Membership on Individuals

Group membership can change individuals, or at least affect the manner in which they behave. Sometimes the change is superficial, but on occasion it represents a major modification of personality and life-style. Some psychologists adhere to the belief that men and women are the product of their individual group affiliations, which shape and mold them accordingly. While the above may be a somewhat extreme view, it seems safe to assume that group membership often causes individuals to conform, to review and sometimes change goals, and to modify their socioeconomic outlooks on life.

Conformity is an essential element in any group, and it acts as a self-perpetuating device. Even people who consider themselves individuals must conform if they hope to enter and remain in a group. The question that each individual must answer is not whether to conform, but how far to conform and how much to compromise. If the group demands too much conformity, as organizations occasionally do, then members will often form informal relationships aimed at lessening the impact of forced formal group affiliations. Sometimes these informal relationships will evolve into a semiformal association such as a fraternal group or even a labor union.

Probably the single most important factor in group membership is personal goal modification, which can mean new values, fresh expectations, and a clearer—or sometimes muddier—personal identity. In some cases, the responsibility that goes with group affiliation may keep original goals from coming to fruition due to physical, economic, and philosophical constraints forced upon group members. Many individuals feel totally "trapped" by their jobs. Perhaps they have too much time invested to be able to retrain for a new occupation, or they have too great an interest in a pension fund to simply pull up stakes and start a new life with new goals. The multiple pressures of marriage and family, going to school, and working can have a powerful impact on goals. Family responsibilities, job requirements, and education are often interrelated, but they frequently impose conflicting values on an individual. In this situation, goal resolution becomes increasingly difficult for the individual, and this can lead to extreme frustration.

A strong argument can be made for the proposition that a person's socioeconomic outlook will be at least somewhat consistent with that of his or her most important group. In many organizations, employee groups are attempting increasingly to represent the political, social, and economic interests of members because the formal organization does not. New members who are inducted into police unions or benevolent associations may not have a firm socioeconomic philosophy when they enter, but their outlook often crystalizes as they interact within the group.

Analysis and Management of Groups

The Activity-Interaction-Sentiment Model

Both George C. Homans and William F. Whyte have written extensively about the elements of group behavior, especially the element of emerging group behavior.[3] Emerging group behavior is that behavior which emerges above and beyond that which is required by external forces to get the job done. Homans and Whyte have identified three basic ingredients for emerging behavior: activities, interactions, and sentiments. An activity is simply something that the group or individual does. An interaction is a communication between persons, either verbal or non-verbal. A sentiment is an idea, a belief, or a feeling that an individual holds. A synonym for sentiment might be an opinion or attitude—it is a liking or preference.

In all groups, individuals perform activities, some required and some not required, or emergent. Individuals are constantly communicating with each other (interactions) about their sentiments concerning the activities. Homans and Whyte suggest that there is a mutual dependency among the three elements of activities, interactions, and sentiments, and that it might work in the following way. Assume that the sentiments that people have can be either positive or negative. For example: One individual might enjoy smoking, a positive attitude, whereas another becomes physically ill at the very smell of cigarette smoke and thus holds a very negative attitude. This illustration could apply to attitudes toward politics, religion, baseball, motherhood, or anything else. If these two persons, one with a positive attitude and one with a negative attitude, are placed in a work situation where both are performing similar activities, it is probable that their interactions will be held to a bare minimum because of the differences in their sentiments. If, on the other hand, two persons are placed together who start out with neutral sentiments, it follows that their association, as they perform similar activities, will probably result in increased interactions, which may then develop into a positive sharing of sentiments. So the relationship between the three elements is a two-way street. The activities of two people affect interactions, and the interactions are affected by the activities as well as the sentiments that each person holds.

The Exchange Theory

Another scheme for analyzing groups is given by Homans and has been classified as a kind of exchange theory. Homans sees groups as always having a purpose—that is, they exist to satisfy certain needs. He believes it is possible to describe behavior in terms of economic concepts such as value, cost, profit, and marginal utility. According to Homans, it is possible to view the behavior between two people as activities that are valued but achieved at some cost. The surplus of reward over cost yields the profit, but an insufficiency may produce a deficit or loss.

For example, assume that we are thinking of two individuals in a police agency. Charlie is the sergeant and Carol is a patrol officer working with him on the day watch. Charlie will reward Carol for performance up to or above the standards required by the department, but the reward obviously could cost money, which may not be provided for by civil service. On the other hand, Carol will reward Charlie for the compensation she receives by producing so many traffic citations, burglary clearances, etc. Keep in mind that there are many nonmonetary and social rewards that can be used in place of money. These include such things as acceptance, recognition, and pride. Since the civil service doesn't provide for monetary rewards beyond the normal salary schedule, Charlie can reward Carol by being more pleasant to her, giving her a feeling of approval and acceptance. In an exceptional instance, he might write a letter of commendation for Carol. This effort costs Charlie in time and thought, but Carol may well respond and reward Charlie for the effort by showing greater loyalty to the agency, greater cooperation, and more interest in the job. Obviously this costs Carol effort. If the reward outweighs the efforts, both Charlie and Carol have a psychic profit.

The same concept can be applied to peer relationships; it is not dependent upon a superior-subordinate relationship. Thus, the activity-interaction-sentiment model plus the exchange theory can give a supervisor some insight into the ways in which groups tends to behave, particularly in terms of emergent behavior.

Dealing with Formal and Informal Groups

One of the problems that a supervisor faces in dealing with his or her work group is the relationship between formal and informal leadership. Some supervisors take the attitude that there cannot be a peaceful coexistence between the two and thus force themselves into a conflict situation. However, the formally designated leader has the option of focusing on one or more of the group objections, for he or she also has personal career objectives. The leader may attempt to be a task-oriented supervisor, directing emphasis, behavior, and sanctions to meet the achievement need. (Most police supervisors probably fall into this category.) If the leader's work group has the same kind of achievement need, there will probably be a perfect match in the productivity and effectiveness of the group. However, in most cases the needs of the formal leader are not perfectly matched with those of the group, and what usually happens is that the group develops emergent behavior to compensate for the deficiency of the formal leader. It is in this set of circumstances that the group will develop an emergent, or informal, leader. This leader will be someone in the group who will give its members direction, identify the appropriate sanctions, or, in some instances, help in the creation of group norms.

Harmony and balance will result in a work group if the formal leader is emphasizing behavior for one basic need while the informal leader is stressing behavior for the other need, but confusion will exist if both leaders are vying for the same kind of objective. For example, if the formal leader is task oriented and the informal leader is oriented toward the social needs of the group, a balance exists and harmony will probably result. However, if both leaders are giving orders and directions for productivity, there is a strong probability of conflict. Similarly, if both leaders are overly concerned about the social needs of the group at the expense of getting the job done, no balance exists and dissatisfaction is likely to follow.

A supervisor is foolish to believe that he or she can predict and control the informal and emerging behavior of the group. The supervisor should never conclude that groups should not be allowed to have an existence of their own or that the words "unofficial" and "informal" are somehow dirty, not to be allowed in the unit. The unofficial, informal, emerging aspects of group behavior are valuable and sound features of any work group that can be turned into a valuable supervisory tool. A knowledge of the relationship between interactions, activities, and sentiments within the group can be of very great assistance to the supervisor when making decisions or recommendations regarding staffing and structure in the agency. The really perceptive police supervisor should recognize personal limitations and perhaps understand that as a formally designated leader he or she cannot satisfy all the needs of the work group. If supervisors do recognize personal limitations, they might encourage the development of informal and emergent leadership so that the balance exists between formal and informal leadership. Another approach would be to use the criterion of *effective work groups* as the supervisor's leadership goal. It might pay to examine carefully some of the excellent research that has been done on work group behavior and the methods of achieving the most effective work groups. Rensis Likert of the University of Michigan identifies the "nature" of highly effective work groups and isolates at least twenty-four separate properties and performance characteristics of the ideal highly effective group.[4] Among the items mentioned by Likert are factors such as behavior that produces a supportive atmosphere and relationship, and a great deal of confidence and trust between the formal leader and the group and also between the members of the group themselves, so that the concept of "our group" rather than that of "my group" predominates. Fear and distrust are removed so that there can be very open and frank communication by all members of the group. The values of each person, of the group, and of the leader are all moving toward balance, harmony, and agreement.

Of course, the Likert model represents an ideal which is seldom fully achieved, but it is the kind of model that a supervisor in a police agency might want to try to achieve. It seems to be a very free, open, and relaxed group, but at the same time, one whose members are aware of the reason for its existence and conscious of the expectations that are being held for it by the organization within which it exists.

Status as a Management Tool

Management in all police organizations makes at least two decisions that, in time, affect the behavior of work groups within the agency. Someone decides on the design of the work flow—the technology of the organization. The second decision deals with the authority system—who has authority over whom, who is to do what, what the rewards and punishments are, how they are to be administered, and what the rules are for employees. Often, these things are set up in a departmental organization chart and manual or book of general orders. The technology factors are important to the development of both group and individual behavior. The authority system defines the roles that are to be designated throughout the agency, and also defines the amount of status that is to be accorded to the various roles. Status and status symbols are a part of organizational life—including life in police organizations—that can have a marked effect on the behavior of individuals and groups. Status may be achieved either from the organization or from other sources. Thus, it becomes an important tool for the supervisor who wants to effect behavioral change through the organizational structure. Almost every police agency has some assignments for its personnel that have a high status value within the agency and other assignments that are considered "punishment details." These may vary from one agency to the next, but they exist in most police and sheriff's departments. It can be seen that the effective supervisor can utilize both leadership skills and the organizational structure in developing a plan of action for dealing with employee group behavior and individual behavior as well.

People can receive status from one of two sources: from the organization or from others. For an individual to have status, others must value his or her attributes, including things such as education, age, sex, training, experience, and seniority. The police supervisor can definitely affect the status of an individual in a work group in terms of the assignment given the person, or in terms of influencing how the other members of the work group perceive the attributes of the individual.

So far, this chapter has tried to convey a number of ideas about individual behavior in groups, how individuals make up groups, and how groups relate to individuals, including the supervisor. Three propositions now suggest themselves:

> 1. First- and second-level police supervisors spend much of their time dealing with groups. The higher the level is, the fewer subordinates there are and the smaller the group is. Also, the higher the level is, the higher are the compensation, authority, and responsibility. Thus, there is a reduced need for control and tight supervision of subordinates in higher-level groups.

> 2. All organizations have informal work groups. Supervisors in formal leadership positions cannot possibly satisfy all the needs of individuals, so informal groups emerge to meet these needs. Supervisors generally are more effective if they maintain a certain "psychological distance" from their subordinates; therefore, groups form to take up the slack in relationships.

3. An organization with strong informal groups will be more effective than an organization with weak informal groups. Needs that are not met by either formal or informal groups will create frustration for the individual and will disrupt organizational objectives and goals. Strong informal groups frequently handle important information faster than the formal channels of communication, although in some cases the information may be distorted during transmission.

Police Employee Organizations

Now it is time to turn our attention to a different sort of group that is part of the daily life of most police officers and which can bear a strong influence on supervision and management within the police agency: the police employee organization.

During 1975 a number of strikes by police officers throughout the United States shook the American public in a manner that will have long-lasting effects on police and community relations. Police officers in San Francisco, Tucson, and Oklahoma City, as well as in a number of other communities, did the unthinkable—they went on strike for higher wages and other benefits. However, the actions they took were not as unique as many members of the public might believe. There have been strikes by policemen many times before, with varying results. It is interesting to note that when the San Francisco officers ran into an impasse in their negotiations with the board of supervisors and went out on strike, the mayor, Joseph Alioto, publicly stated that any police officers who did not return to work immediately would be fired. The strike continued, and eventually the mayor invoked his emergency powers to override the decision of the board of supervisors and gave the police department nearly all of the wages and other benefits they had been demanding. No one was fired. However, in the next city election several proposals that were being sponsored by the police were soundly defeated by the voters, and most observers attributed this to public indignation rising out of the strike.

The phenomenon of greater militancy on the part of organizations representing police officers is one which has been growing steadily for the past ten or more years. A number of police organizations have been gaining more power nationwide and are becoming much more vocal in their demands for better wages and working conditions for their memberships. Organizations such as the Fraternal Order of Police (FOP), Peace Officers Research Association of California (PORAC), the International Brotherhood of Teamsters, Chauffeurs, Warehousemen and Helpers, the American Federation of State, County, and Municipal Employees (AFL-CIO), various Police Benevolent Associations, and others have all been extremely active in recruiting new police membership during the past several years. With the advent of a new push in a number of state legislatures for collective bargaining for public employees, including police, it is obvious that the unions and related organizations will continue to press hard for increased membership and bargaining rights.

International Police Unions

Of interest is the fact that in most European countries, police unionism is already highly developed.[5] Virtually all police officers in the Scandinavian countries, the Federal Republic of Germany, France, England, Switzerland, and other areas of Europe belong to a police union or association. On the other hand, the unionization of police has not been found necessary or, for that matter, desirable in many African and Asian countries, including Japan and Pakistan where there is a continuing liaison between employee and employer for the purpose of improving salaries and working conditions in the police service. In the United States, it appears that most members of large police forces maintain membership in some form of police union or association, although the structure of these units and their role in bargaining for employment conditions may vary in different areas.

A number of European police administrators as well as rank and file officers were interviewed by one of the authors during the summer of 1975 and it was noted that they were all in agreement on one point—that is, a strike by the police is unthinkable. A patrolman in Switzerland went so far as to say, "Any police officer who would even think about going on strike should change his occupation immediately. He has no business being a policeman." A German detective stated that his union confines itself to working constantly to improve economic conditions for the officers and would not under any circumstances intervene in a disciplinary action involving a member except to provide him with an attorney to represent him in hearings or other actions.

The situation in New Zealand, which has a national police system, is interesting from the American viewpoint. Almost all constables, sergeants, and senior sergeants are represented by the New Zealand Police Association. About 99% of the personnel in these ranks belong to the association, though membership is not compulsory. The fees run about $20.00 per year for membership. Inspectors and all ranks above (except the commissioner) belong to the New Zealand Police Officers Guild. Fees are about the same as for the association and, again, 99% of the commissioned officers are members.

Wages for police officers normally are set each year by automatic cost of living adjustments along with those of all other civil servants. If there arises a need to realign police salaries with comparable occupations in private employment, this is accomplished by negotiations with the Treasury Department and through the Minister of Police with the New Zealand Cabinet. Sometimes subtle pressures are brought to bear through friendly members of parliament and the news media. Occasionally, the Association and the Guild will use the services of private consulting firms to do economic comparisons with other organizations and occupations.

The terms of employment are set forth in the Police Regulations and are based upon representations made by the commissioner and the two service organizations to the Cabinet. The salary schedule is revised each year and sometimes twice a year, depending upon government policy at the time.

Grievances, if not resolved at the supervisory level, may be submitted in writing to the commissioner and, if it is desired, under confidential cover. (Confidential cover means that it *must* be forwarded by higher ranking officers.) Each officer above the complaining officer in the chain of command is required to make a recommendation with regard to the grievance. The officer in charge of the district is required to make whatever inquiries are necessary and then report to the commissioner as to whether or not, in the officer's opinion, the complaint is justified. If the officer finds that the complaint is justified, this person can take the action necessary to rectify the situation. In the event that the complaint is not justified, it goes on up to the commissioner for final resolution.

The Association and Guild do not involve themselves in internal disciplinary actions except to provide counsel to the accused and pay expenses if they feel that the officer was acting in good faith.

The commissioner has the power to make some minor salary adjustments, within the limits of the budget. Usually, these are in relation to allowances for items such as plainclothes, travel expenses, mileage, etc.

Trade unions and guilds have been a very potent factor in New Zealand politics for many years, and no one sees anything at all unusual in the fact that the police should have such organizations representing them.

The Development of American Police Unions

The development of labor unions in the United States in private enterprise resulted largely from the working and living conditions that were created by the industrial revolution.[6] Police employee organizations were also a response to social conditions, but most of their early activities did not appear to be economically based. The first rank and file police organizations began to appear toward the end of the nineteenth century. The labor movement was characterized by great upheavals and continuous protests against working conditions. Often violence erupted in labor disputes, and too often the police were thrust into the center of the controversy as strike-breakers, which brought about a good deal of resentment between workers and policemen. Naturally, this created a sense of alienation between the police and the community. The police reaction was predictable; they started to gather themselves into social, fraternal, and benevolent organizations. The craft unions in particular gained strength during the period from 1900 to 1930 and each confrontation between management and labor seemed to leave labor in slightly better condition than it had been in before.[7] Needless to say, policemen, who were still caught in the middle of these confrontations, began to notice the gains being made by labor while policing was still relegated to a minor situation by most governmental entities. During the 1920s, policemen began to unionize because of their rapidly deteriorating economic status. Most unions during this period had rules forbidding police membership, so there was practically no success in affiliating with recognized trade unions. However, local police associations continued to develop and a number of state and national police organizations began to emerge and gain recognition.

Police Organizations Today

Most police officers today belong to some form of employee organization which, while not often officially classed as a trade union, shows a surprising degree of correspondence to a conventional union. In general, we can identify such organizations as local, state, or national police organizations, which often maintain direct ties between the various levels.

Local Organizations

The local police organization is unquestionably the most important unit in the police labor movement. While state and national organization representatives may give advice and counsel to the officers of the local union, they cannot become too deeply involved in local affairs and politics; they usually have a different role to play. Most funding for police services comes from local taxation and is allocated locally by city or county mayors, managers, and/or governing bodies such as councils and boards of supervisors. Police services to local taxpayers are largely performed by local police officers. Most of the pressures which a union might bring to bear to affect salaries, hours, and other conditions of employment have to be exerted locally through bargaining with the city or county or by engaging in political activities in the local arena.[8] For these reasons the local union has the major responsibility for any activities which are undertaken on behalf of its members.

The local character of the employment relationship not only helps explain why the local union is so important in the police labor movement, but also helps explain why the large, relatively centralized national police organizations have not been able to enroll a really large membership, considering the great numbers of police officers in the United States. Basically, it appears to be mainly a matter of simple economics: Why pay dues to a regional or national labor organization when the essential responsibilities will be carried out by local union leaders interacting with local officials in a local political and collective bargaining situation? The dominant state and national organizations focus their attention on the lobbying and information needs of the local police unions.[9]

State Organizations

State organizations appear to be the next most important level in the police labor movement. Again we find a mostly functional explanation. In many cases, because of provisions in either state constitutions or statutes, some aspects of the employment relationships in cities and counties may be under the jurisdiction of the state legislature. Usually, aspects such as pensions and death and disability benefits are handled by the state. Another issue that is usually resolved by state statute is the entire question of collective bargaining for public employees, sometimes with special provisions for public safety services. Another state-level issue is the definition of crimes and punishments.

Many local organizations affiliate with state-level organizations for the purpose of bringing political pressure on state legislatures. Usually, state organizations don't directly represent local police in collective bargaining negotiations. Rather, as a state group they have more political "clout" in their lobbying activities. They also collect and disseminate wage and hours information and provide legal, financial, and informational assistance to their member associations.[10]

National Organizations

The largest of the national police employee organizations is the International Conference of Police Associations (ICPA), which in 1972 claimed more than one hundred local and state police organizations representing a total of 158,000 police officers. As the name implies, membership is organizational rather than individual and the membership is concentrated in California, New York, Illinois, and New Jersey. A number of the largest city unions are members, including those in New York, San Francisco, Chicago, Washington, D.C., St. Louis, and Seattle.[11] ICPA has, in recent years, become quite active and maintains an ongoing and regular lobbying program in Congress.

The second largest of the national groups is the Fraternal Order of Police (FOP), with 733 affiliated lodges representing about eighty thousand members. The majority of the FOP membership appears to be concentrated in Pennsylvania, Ohio, Florida, Illinois, and Indiana. While the ICPA was made up of a group of existing state and local associations and admits other associations which apply for membership, all of the local, state, and national units of the FOP have the same name, and the national lodge of the FOP issues charters to local lodges in much the same manner as national labor unions issue charters to their locals. Additionally, an individual police officer may hold membership in the national FOP lodge as well as being a member of a local lodge. One of the important regular functions of the national FOP has been the publication of an annual salary and working conditions survey. During the 1971 convention, the delegates voted to delete the constitutional prohibition on political activity. By contrast with its earlier efforts to remain completely disassociated from organized labor, in recent years the FOP has become very much involved in obtaining collective bargaining rights for police officers.[12]

Other major national organizations that represent police officers are: the American Federation of State, County, and Municipal Employees (AFSCME); the National Union of Police Officers (NUPO), which is affiliated with the Service Employees International Union (SEIU); and the International Brotherhood of Police Officers (IBPO), which is affiliated with the independent National Association of Government Employees (NAGE). The Teamsters Union has also made sporadic efforts to organize police locals but, to date, has a comparatively small membership among police officers.[13]

It is easy to see that there is a large number of diverse types of police organizations in existence today. Bopp states that "it is becoming increasingly difficult to articulate an acceptable typology of police employee organizations by classifying them according to name."[14] He goes on to classify them by functional type and characterizes local police employee organizations as follows: (1) the camaraderie society, (2) the professional association, (3) the pragmatically militant union, and (4) the subversively militant union.[15]

It appears that most police employee organizations fall, to a great extent, into one of the above categories or are in a state of transition from one to another. Future developments in the field will determine whether or not we can continue to use Bopp's functional typology. The movement toward police unionization is still in its infancy, and it is probable that there will be many changes in the not-too-distant future.

Police Militancy

During the middle and late 1960s, four general factors seemed to make major contributions to dissatisfaction on the part of police personnel. The first of these was an obvious increase in hostility from the public. Minority and student militancy, increased public interest in the creation of civilian review boards, adverse Supreme Court decisions, and perceptible nonsupport from local officials can all be cited as evidence of the public's attitude.

The second factor was the "law and order issue"—an increasing demand by the public for personal safety in the face of rising crime rates. Thus, some of the same environmental factors—such as poor housing and poverty—which made the officer's job more difficult also created increased demand for more effective police work. Since it should be obvious that the police cannot, by themselves, completely control or prevent crime, these external factors have put the police in a situation where they cannot win no matter what they do.

This problem was further aggravated by the third factor—a growing recognition by many officers that their pay scales were lagging far behind the pay and benefits enjoyed by other public employees as well as most employees with similar qualifications in private enterprise.

The final factor helping to explain police dissatisfaction was—and still is—the widespread existence of poor personnel practices in police agencies. Abuses that are still common today are: lack of civil and constitutional rights for officers who are under investigation themselves; lack of a functional grievance procedure; no pay for off-duty court appearances; no premium pay for overtime work; arbitrary transfers; favoritism in promotions and assignments; and others.[16]

While the four factors listed above suggest a possible explanation for increasing police militancy, three other factors may well be important in contributing to this new militancy: the success of other groups in achieving their goals through militancy, the influx of young officers into departments, and the high degree of group cohesiveness found among police officers.

However, in the long run the need for personal development of power and identity may prove to be more important to officers than wages and working conditions. Officers need to develop a sense of self-worth before a client-service orientation can really emerge. The unrest of the sixties made the police more aware of their own position in society. They began to question their place in the American social structure and suddenly became much more aware than they had been in the past of who their friends and enemies were. They experienced a general feeling of powerlessness. They suddenly realized that they had no real aims and goals, were lacking benefits enjoyed by other workers, had no real job mobility, and were being exploited by politicians in their rhetoric. From this type of mental state emerged the militancy we see today in the American police.

Labor Relations and the Police

Police unions operate within a statutory context which is complex and often confusing. There are multitudes of local, state, and federal laws which govern the conduct of labor relations for public employees and, in some cases, there are special provisions governing public safety services. Militant behavior on the part of police unions really entered the scene during the late 1960s and it has been with us since. All police employees have the right to join an employee organization of their own choice under the provisions of the First and Fourteenth Amendments to the U.S. Constitution. However, only twenty-seven states, as of 1974, had legislation that makes it mandatory to recognize and bargain with the representative of the employees. Unlike the private sector, which for the most part is regulated by federal labor law, labor-management relations for municipal and county employees are regulated mainly by state legislation. No jurisdiction allows police officers to go on strike, but on the other hand, no jurisdiction has successfully legislated against strikes.[17]

Negotiation Techniques

In an effort to eliminate strikes, states have tried a variety of techniques to resolve impasses in contract negotiations. These techniques are discussed below.

Mediation, which is a noncoercive process in which a neutral third party attempts to work behind the scenes to help the parties reach some sort of agreement. He or she does this by maintaining the channels of communication, by persuading and cajoling, and by offering suggestions for solutions. If the mediator is not successful in bringing the parties together in agreement, a strike may follow or another alternative may be tried.

Fact finding, which is a procedure in which a neutral third party holds a hearing and, after consideration of all the facts presented, makes recommendations as to what the new terms and conditions of employment should be. However, these recommendations are not binding on either party to the dispute. Resolution of the dispute is largely dependent on the concept that the public will accept the fact finder's neutrality and support his or her findings, thereby putting a good deal of pressure on both parties to accept the findings. In the use of this procedure, however, reality sometimes differs from theory.

Voluntary binding arbitration, which is essentially fact finding with a prior agreement by the parties to accept all or part of the award as binding. This technique has been quite effective except in those situations where one party or the other cannot or will not voluntarily agree to accept the binding aspect of the process. Usually, the reasons for nonacceptance are primarily political in nature.

Compulsory binding arbitration, which requires arbitration at the request of either party and makes the ruling binding on both parties. While there is no question that compulsory arbitration does prevent strikes, there are several major problems with it. First, the parties may not move from their initial positions if they are reasonably sure that an arbitrator will split the difference at a later date. For this reason, there is little incentive to engage in the give-and-take of good-faith negotiations. Second, the judgment of the arbitrator regarding what constitutes an acceptable package is substituted for the bargaining process. Third, and probably most important in the public service, is the fear of governing bodies of cities and counties that the arbitration award would be in excess of what the tax base could handle. Fear has been expressed in many quarters that passage of legislation mandating compulsory binding arbitration could lead a number of local governments into bankruptcy or insolvency. For this reason, a great deal of lobbying effort has been expended against such legislation at the state level. Another argument that has been advanced against binding arbitration is that it puts into the hands of a single individual decision making which has been delegated by the public to its elected representatives.

All of these arguments notwithstanding, until another acceptable substitute is presented, binding arbitration appears to be the most likely means of avoiding strikes by public service employees. John Burpo, who is quite well known as a labor authority in the police sector, has called most of management's objection to binding arbitration fallacious. He describes this type of arbitration as "the ultimate strike alternative."[18] Legislation appears to be the only logical means of implementing a policy of binding arbitration, and a number of states and communities have set up such statutes.

Job Actions

The first police strike in American history which can be verified took place in Ithaca, New York, in 1889. The strike, which lasted for a week, involved five officers who were protesting a cut in their pay. Probably the best known of the early police strikes was the 1919 strike by officers in Boston, where the main issue was the right to unionize. This strike was a total disaster for the police; more than eleven hundred of the strikers were fired and barred from ever coming back to work in the police department. For over forty years after the Boston strike, American police gave little indication that they would ever again be involved in strike action, but the militancy of the 1960s changed the picture completely.

Police employee organizations have adopted a number of tactics, some legal and some otherwise, that stop short of an actual strike for the purpose of enforcing their demands on government. Examples of some of these tactics follow.

One of the devices most frequently used is that of political action. Police officers have awakened to the fact that they are not second-class citizens, and a number of court decisions have upheld their right to participate in political affairs. One of the principal activities in which police organizations engage is that of political lobbying at all levels. All over the United States one can find full-time paid lobbyists representing police employee groups before legislative bodies. The police have found that it often pays to bypass both their own administrators and public officials by carrying their case to the public via the ballot. Probably the best-known successful referendum was that initiated by the New York City patrol officers to get rid of a much hated civilian review board. Oakland, California, officers have been quite successful in winning benefits by means of the ballot. San Francisco police officers also enjoyed a high success rate, until shortly following the strike mentioned earlier.

We also find police organizations supporting individual political candidates both financially and through endorsements. Individual police officers are now running for public offices ranging from local boards of varying kinds to state legislatures. A number of mayors of large cities are former police officers. Apparently the police officer who aspires to public office can be a highly attractive candidate, if the number of such people being elected is any criterion.

Another tactic gaining favor with police organizations is that of pursuing litigation in the courts. Suits have been filed regarding the payment of overtime, unauthorized payroll deductions, "illegal" department policies or civil service rules, and many other items which have not been resolved to the satisfaction of the rank and file officers. Many of the court decisions have been favorable to the police and have had widespread ramifications for local government.

Administrative disobedience is another technique that has been adopted in recent years. This tactic is one that can cause some really serious problems for a supervisor in day-to-day operations. Supervisors are charged with enforcing agency policy but are finding that the day of blind, unquestioning obedience to all orders is fast leaving the scene. Today's young officer is quite likely to raise questions about orders that appear to be inadequate, inappropriate, unjust, unworkable, unsafe, discriminatory, or just stupid. A strong union seems to make the use of this particular tactic more common. One of the most common areas of disobedience seems to be that of limitations on the use of firearms that go beyond what the law requires. This also extends into the area of weaponry and ammunition when the officers feel that they are "undergunned" in comparison to the criminal element. It is extremely difficult to bring disciplinary action when great masses of personnel refuse to adhere to an unpopular policy. This is particularly true when the supervisors also are not supportive of the policy. A related but different tactic is the use of the "vote of no confidence" by the rank and file personnel; this is usually directed at the chief of police. It is embarrassing to the chief and places him or her in a difficult position when it comes to explaining the event to the governing body of the community and to the media. There are several instances where this tactic has led to the resignation or early retirement of a chief, but in more cases than not the chief has managed to ride it out until the situation quieted itself or an acceptable compromise was reached.

Other tactics short of a strike are mostly work related and include such things as slowdowns and speedups. Many of these actions involve traffic citations. The officers may simply stop issuing routine citations until they get what they are after. Imagine the effect of total lack of traffic enforcement in the congested business area of a major city. A tactic opposite in method to the slowdown is the issuance of citations for every technical violation observed. This too will create an awareness of the problem on the part of the public as well as pressure for quick settlement of the issues. In some cases where a restraining order forbidding a strike has been issued by a court, there is a sudden epidemic of "blue flu." Large numbers of officers call in sick and don't get well again until the issues have been settled.

Labor Relations and the Police Supervisor

From the foregoing discussion it should be readily apparent that the supervisor needs to be fully cognizant of the many ramifications and implications for the job that are inherent in the growth of employee organizations and their increasing militancy. It is essential that the supervisor be completely aware of the pertinent provisions of any contract governing working conditions, employee benefits, and particularly grievance procedures. If unaware of contract provisions, the supervisor could easily and inadvertently create an incident that might lead to a grievance and personal embarrassment, as well as embarrassment for the administration of the department. Some actions supervisors could have taken with impunity in the past are no longer allowable under the contract, and it may be necessary to alter the style of operation. Both supervisors and police managers are going to have to accept the idea that decision making with regard to departmental policies can no longer be unilateral. Consultative procedures are often required under contracts, and there will have to be a sharing of power that may not have existed in the past within the agency. The idea of giving up powers that have traditionally been a sole prerogative of supervisors and managers meets with a great deal of resistance from traditionalists in the field. However, it must be pointed out that business and industry long ago learned to live with labor unions, and while the relationship has had its ups an downs, by and large there have been many mutual benefits. The same evolutionary process will have to take place in the police service. It is safe to predict that police unions will eventually become an accepted part of the public safety service scene and that both management and unions will continue to mature until a normal and harmonious relationship develops. The supervisor is the key person in helping to create in his or her unit a working climate that will lead to harmonious labor relations and an effective work group.

DISCUSSION QUESTIONS

1. Do all formal groups have informal groups? Explain.
2. How do loyalties to groups develop?
3. Do group memberships influence individual behavior? How?

4. Define a group. Why is the concept of the group so important to supervision?
5. What are the characteristics of a formal work group? Of an informal work group? Of the semiformal group?
6. What characteristics and processes are common to all groups?
7. What do group members expect from the group?
8. What do groups expect of their members?
9. What are the principal effects of group membership on individuals?
10. What is emerging behavior relative to the group process?
11. What are some of the ways in which a group can be analyzed?
12. How can the supervisor utilize informal group leadership to best advantage?
13. What are some of the characteristics of highly effective work groups?
14. What is the role of status and how is it achieved?
15. How can the supervisor use the need for status to effect behavioral change?
16. What are the different levels of police employee organizations?
17. How do you account for the marked increase in police militancy?
18. Contrast the role of police unions in the United States with the role of unions in other countries.
19. How do the different levels of police organizations tend to function?
20. What are some of the processes used to avoid strikes in the settlement of labor disputes?
21. Discuss the potential problems for the supervisor that are inherent in labor contracts.

NOTES

1. Joseph L. Massie and John Douglas, *Managing: A Contemporary Introduction* (Englewood Cliffs, N.J.: Prentice-Hall, Inc., 1973), p. 85.
2. See Elton T. Reeves, *The Dynamics of Group Behavior* (New York: American Management Association, Inc., 1970).
3. See George C. Homans, *The Human Group* (New York: Harcourt, Brace and Co., 1950); George C. Homans, *Social Behavior: Its Elementary Forms* (New York: Harcourt, Brace and World, 1961); and William F. Whyte, *Men at Work* (Homewood, Ill.: Richard D. Irwin Co., 1961).
4. See Rensis Likert, *New Patterns of Management* (New York: McGraw-Hill Book Co., 1961).
5. "The Emerging Roles of the Police and Other Law Enforcement Agencies" (Working Paper, Fifth United Nations Congress on the Prevention of Crime and the Treatment of Offenders, Geneva, Switzerland, September 1975), p. 29.
6. Arthur M. Schlesinger, Jr., *The Coming of the New Deal* (Boston: Houghton Mifflin Co., 1959), p. 385.

7. William J. Bopp, *Police Personnel Administration* (Boston: Holbrook Press, 1974), p. 322.
8. O. Glenn Stahl and Richard A. Staufenberger, eds., *Police Personnel Administration* (North Scituate, Mass.: Duxbury Press, 1974), p. 210.
9. Ibid., p. 210.
10. Ibid., pp. 210-211.
11. Ibid. p. 211.
12. Ibid., pp. 211-212.
13. Ibid., p. 212.
14. Bopp, op. cit., p. 323.
15. Ibid., p. 323.
16. Stahl and Staufenberger, op. cit., pp. 205-206.
17. See Hervey A. Juris, "Coping with Police Strikes: Some Legislative Alternatives," *Police Yearbook, 1972,* Washington, D.C.: International Association of Chiefs of Police, 1973.
18. John H. Burpo, *The Police Labor Movement* (Springfield, Ill.: Charles C. Thomas, 1971), pp. 137-140.

10 Discipline

Discipline is a function of command that must be exercised in order to develop a force amenable to direction and control. Unfortunately, the word "discipline" has taken on a disagreeable and negative meaning. There is a tendency to think of it entirely in its most limited sense—an action taken against an employee who has been guilty of some violation of good behavior. This is sometimes referred to as "corrective discipline."

Discipline may also be thought of as that force that prompts an individual or group to observe rules, regulations, and procedures that are deemed necessary to the attainment of an objective; as that force, or fear of force, that restrains individuals or groups from doing things that are deemed destructive to group objectives; or as the exercise of restraint or enforcement of penalties for the violation of group regulations.

While this is one use of the word "discipline," there is a broader and more positive meaning. It can be thought of as a form of training and an important, constructive tool of leadership used to eliminate weakness and prevent its reappearance. We speak of employees as being well disciplined or poorly disciplined. The conduct of well-disciplined employees conforms to the employer's standards of behavior. Poorly disciplined employees do not conform to these standards.

The word "discipline" has its roots in the Latin word *disciplina,* which means instruction, teaching, or training. An undisciplined group is one that is incompletely trained, not because of a failure of the formal training program, but because supervisory personnel have not required that subordinates conform to the department's rules and procedures. The best-disciplined police forces are the best trained and, for this reason, the least punished. Well-disciplined employees accept and live according to certain behavior patterns, voluntarily and often without conscious effort or thought. This conduct results when proper working habits are established and maintained over a long period of time. It is enforced by either group discipline or self-discipline. A high morale and *esprit de corps* may develop within a group, and a self-administering group discipline results. In addition to preventing situations that demand corrective discipline, such a group feeling creates within each individual employee the desire and determination to reach the work objective regardless of the obstacles encountered.

Discipline in the Police Service

The problem of discipline is not restricted to police agencies, but because of the police organization's quasi-military nature, personal and group discipline is of greater importance in law enforcement than in other government agencies or in industry. The special requirements of crime suppression give the problem of discipline a particular importance in police work. An officer who has difficulty abiding by the department's regulations is of doubtful value in the enforcement of law.

Characteristics of Poor Discipline

The major characteristics of poor discipline are the following:

1. Low morale

2. Lackadaisical attitude toward job, superiors, the department, and the public

3. Lack of direction or objective

4. Inattention to duty

5. Violations of rules and regulations

6. Disregard for the rights of the public, resulting in the loss of public confidence and trust

Code of Ethics

One of the first requirements of good discipline is a clear, concise code of conduct, or ethics. Many such codes have been written. In California in 1956, a code of ethics was adopted by several police organizations. The code was drafted by a committee of the Peace Officers Research Association of California, edited by a committee of the California Peace Officers Association, and adopted by both organizations. This code reads as follows:

> As a Law Enforcement Officer, my fundamental duty is to serve mankind; to safeguard lives and property; to protect the innocent

against deception, the weak against oppression or intimidation, and the peaceful against violence or disorder; and to respect the Constitutional rights of all men to liberty, equality, and justice.

I will keep my private life unsullied as an example to all; maintain courageous calm in the face of danger, scorn, or ridicule; develop self-restraint; and be constantly mindful of the welfare of others. Honest in thought and deed in both my personal and official life, I will be exemplary in obeying the laws of the land and the regulations of my department. Whatever I see or hear of a confidential nature or that is confided to me in my official capacity will be kept ever secret unless revelation is necessary in the performance of my duty.

I will never act officiously or permit personal feelings, prejudices, animosities, or friendships to influence my decisions. With no compromise for crime and with relentless prosecution of criminals, I will enforce the law courteously and appropriately without fear or favor, malice or ill will, never employing unnecessary force or violence and never accepting gratuities.

I recognize the badge of my office as a symbol of public faith, and I accept it as a public trust to be held so long as I am true to the ethics of police service. I will constantly strive to achieve these objectives and ideals, dedicating myself before God to my chosen profession . . . law enforcement.

The Nature of Discipline

There are three general categories of discipline:

1. Positive discipline, or the force that prompts individuals
2. Negative discipline, or the force that restrains individuals
3. The actual imposing of penalties

Positive Discipline

When discipline is used in its positive sense, it means that a group of people is well trained, well motivated, capable of intelligent cooperation and of operating well without direct supervision, and willing to respond completely to supervision. Positive discipline will be discussed in detail later in this chapter.

Negative Discipline

Negative discipline is the controlling of behavior through punishment, chastisement, or threats. Many experts feel that when negative discipline must be applied, there has been a breakdown of leadership. Others feel that negative discipline is both ineffective and wasteful because the behavior it produces is not willing, sustained, or fully cooperative, i.e., officers cooperate to the degree required to avoid punishment but do not necessarily cooperate toward the goals of the organization. Also, negative discipline appeals only to fear, which is one of mankind's most primitive instincts.

Control

There are three levels of control: fear, promise of reward, and appeal to higher motives. Negative discipline requires *constant* supervision, as men and women respond only when the supervisor is in a position to evaluate their performance, while positive discipline is established by leadership and by observing the human relations aspects of supervision. Negative discipline is necessary in a police department for two reasons:

1. Because of failures in leadership, recruitment, and supervision

2. Because of the nature of police work, the necessity for a scapegoat when the public is exerting pressure, and the various temptations that occur in police work, which may be so strong as to overcome the effects of leadership

The Supervisor's Use of Negative Discipline

The supervisor's position, midway between management and subordinates, creates many painful situations. First, the supervisor is the one who observes or is made aware of the situation that requires negative discipline. Second, the supervisor is the one who must use negative discipline from time to time to maintain his or her position and the work level of the group. Yet it is a gamble every time it is used, because:

1. The supervisor may not be supported by the management and, in that case, will suffer loss of prestige in the eyes of subordinates.

2. The supervisor may apply more severe discipline than the officer deserves and will be condemned as too harsh by subordinates.

In either case, unless the erring officer is not liked by peers or has done an unnecessary deed, the supervisor is incurring a certain amount of hostility from other officers. Hence the supervisor is sorely tempted to neither see nor report actions that will require negative discipline.

The role of the supervisor has been constantly reduced in the area of discipline, and the authority of the supervisor is, therefore, minimal and tightly controlled by law, rules, and management. Usually the supervisor can only relieve an officer from duty without management's approval; he or she cannot suspend or dismiss. In some departments a supervisor cannot even relieve an employee from duty, but must place the individual on "locker room" duty, pending management's approval. Yet the supervisor must continue to observe, gather data, write incident reports, etc., or the whole system breaks down. When positive discipline breaks down, negative discipline soon follows.

Negative discipline is the most unpleasant supervisory duty. It is shirked by some supervisors, but never by a *leader*. The supervisor's role in the negative discipline system requires that he or she:

1. Must observe the officers' performance.

2. Must investigate complaints against officers.

3. Must report violations observed and cases investigated. (In some instances, continual note taking on a situation and the actual preparation of a case are part of this role.)

4. Is expected to testify, if necessary, at board hearings.

5. Must administer certain types of disciplinary action, when authorized, such as an official oral reprimand.

Types of Negative Discipline

Informal Type

This group system of discipline requires the cooperation of several of the offender's fellow officers, and it can be used only by a supervisor who has developed a rapport with most subordinates. This type of discipline is just the old silent treatment. The group deprives the offender of companionship until the person's conduct conforms with accepted standards. This is a very effective type of discipline when conditions are right.

Formal Types

Formal types of negative discipline include the following, listed in order of severity:

1. Reprimands
2. Transfer
3. Fine or extra duty
4. Suspension
5. Demotion

6. Dismissal

7. Criminal action

Evaluation of the Offense

Punitive or corrective action is necessary when rules and regulations are violated; the question is, how much and what type. When recommending punishment, the supervisor must make important judgments about the motive and intent of the offender. The supervisor first must decide whether the violation occurred as a result of deliberate defiance of the rules and regulations, or inadvertently as a result of ignorance or carelessness. If the latter was the case, punishment should be directed at retraining the officer, helping the officer to increase his or her value to the department. However, if the violation was an act of defiance, immorality, or dishonesty, the punishment should be severe. Employees expect and want uniform adherence to recognized standards of conduct, and they respect the superior who maintains these standards. When corrective disciplinary action seems necessary, the supervisor should take the following action: (1) The facts should be obtained promptly; and (2) Action should be taken as soon as possible so that the employee's action and the supervisor's corrective action come close together. This does not mean that the supervisor should act before the facts have been gathered and weighed. It *does* mean that the supervisor should act as soon as he or she has acquired all the facts, including the employee's side of the story, has weighed the evidence, and has decided what to do on the basis of the facts. The longer the corrective action is delayed, the more unjustified and unfair it will seem to the employee and his or her fellow officers.

In deciding what corrective disciplinary action to take or recommend, the supervisor should consider the following:

1. All the circumstances surrounding the case

2. The seriousness of the officer's conduct in relation to his or her particular job

3. What the department has done to help prevent this type of behavior

4. The corrective disciplinary actions suggested in the department's manual for the type of offense involved

5. Any corrective action that would have a training value rather than a strictly punishment action in reprisal for the offense

6. The disciplinary actions of a corrective nature the department has taken in similar instances

7. The officer's previous conduct record in the department

8. The probable cause of the officer's behavior

9. What corrective action will most likely eliminate the cause of the offense and prevent a recurrence

10. The officer's probable reaction to the corrective action

11. The possible reactions of the other officers to the corrective action

The three primary purposes of punitive action are: (1) to punish an employee in order to produce acceptable conduct, (2) to eliminate an employee who is, or has become, unsatisfactory, and (3) to serve as an example.

Punishment designed to reform the conduct of an officer generally is warranted in two types of cases: (1) a single offense or series of closely related offenses that do not warrant dismissal when the offender's general record has been satisfactory; and (2) a series of minor infractions over a period of time that show an officer's unwillingness to conform when a reprimand has not produced improvement.

The dismissal of an officer may be necessary in the same type of cases with these additional circumstances: (1) a single offense or series of related offenses of such a nature that dismissal is necessary in order to maintain the respect and confidence of the public, establishing that such offenses will not be tolerated, and permitting other employees to maintain their respect in the department, and (2) when a series of minor infractions have occurred over a period of time and all other means of correcting the officer's conduct have failed.

Disciplinary Transfers

Occasionally, the supervisor finds it necessary to impose a disciplinary transfer on an employee who has not responded to other disciplinary action. This type of transfer should be considered carefully before being used. A morale problem is created if employees get the impression that a "Siberia" has been found for negative employees. Such an impression can be avoided by judicious use of the disciplinary transfer.

The purpose of the disciplinary transfer is to move an employee to a place where management will be able to provide the proper level and amount of supervision and ensure an effective evaluation of the individual's efforts and performance. This evaluation should determine if the employee can be retained in the service without reducing the effectiveness of the organization. The disciplinary transfer should always include a preset time schedule for evaluation which both the employee and the supervisor understand.

A written record of evaluations of the employee and consultations with him or her must be kept. Action should be taken as soon as it is evident that the employee is, or is not, demonstrating a desire to conform to regulations. Whether or not to continue observations should be determined by the employee's behavior. Documentation is always required in dealing with potentially useful employees who have gotten off the track. Periodic reviews should keep all levels of supervisory personnel informed of the progress made and the techniques used.

When the employee's behavior changes for the better, the employee should be restored to full normal-duty status. If the employee cannot conform to the standards expected and adjust to the situation, action should be taken to remove the individual from the service. The documentation made during the attempt to restore the employee to full status will show what was done to protect the interests of both management and the employee, and will serve as a good example of professional management technique.

Transfers of a disciplinary nature should never be used as a form of harassment to force an employee to submit to the "yoke" of management. The misuse of this technique might bring about the opposite of what was intended: grievances or low morale. The disciplined employee will be quick to point out any ill-advised act of management as an example of the suffering the person is enduring as a "misunderstood employee."

Shortcomings of Punitive Action

The punished officer rarely responds to leadership, usually suffers from a negative attitude, and rarely is "organization" oriented. Punitive action sets precedent both as to *act* and *extent*. The supervisor may wish to deviate from this procedure in the future, since like acts do not always demand like punishment. However, differences in situations are rarely appreciated at the police officer level.

Hearings

After a department announces the discipline to be administered, the offending officer generally has the *right* to a hearing or an appeal. This right depends on the organization and the offense.

Regardless of what the system may be called, generally there is an established set of rules that governs the hearings. These rules usually cover two broad areas that concern management and employees:

> 1. Rules that define the powers of the commission and the departments, and the various technicalities which apply to these powers
>
> 2. Rules that define the obligations, rights, or privileges of the individual employee

The right of appeal is our immediate concern. The types of cases that generally require a formal presentation to the commission or board, if the appeal privilege is exercised or provided, fall into several categories:

1. Separation of a probationary employee from the service, unless the appointing authority has the final word

2. Separation of a permanent employee from the service.

3. Reduction in rank of a permanently-appointed individual.

Generally speaking, all other forms of negative disciplinary action are the prerogative of the department head, and the employee does not have the right of appeal. Examples of such cases are the following:

1. Suspension without pay. The rules, however, generally limit the length of suspension which the department head can impose.

2. Oral and written reprimands.

There is generally a right of review in these less severe forms of disciplinary action, but usually there is no right of formal appeal.

The factors that are necessary to discharge a probationary employee vary with different jurisdictions; however, some of the most prevalent ones are the following:

1. Evidence that the employee has received adequate counseling concerning mistakes or shortcomings and that the employee has failed to improve satisfactorily. Adequate counseling is a variable factor, but a good rule of thumb is that it consists of at least three or four counseling sessions in a six-month period and preferably one counseling session for each month of probation. In essence, the probationary employee should be made aware of the department's opinion of his or her performance. Employee evaluation reports should include data on counseling sessions and should specify areas of weakness.

2. Evidence that the employee has committed an offense serious enough to warrant discharge even if he or she is a permanent employee. An example of such an offense is an arrest for drunken driving.

When discharging a probationary employee, a supervisor should submit a written statement of facts outlining the employee's failure to meet minimal standards and a copy of the official form used for grading the employee's performance. Most agencies keep either monthly or quarterly performance evaluation files on probationary employees.

It is customary in many agencies to tell a probationary employee that his or her work is not satisfactory and that the employee is to be discharged at the end of probation. This gives the employee an opportunity to resign before the end of the period. In honoring this custom, the supervisor must remember three things: (1) If the employee is to be fired for serious misconduct, or violation of the law, due to the possibility of compromising the incident the employee should not be given the opportunity to resign. (2) If an employee is given the opportunity to resign, the employee must take advantage of it *before* the probationary period is up. (3) The supervisor must be extremely careful not to "offer" an employee the opportunity to resign when the employee is, in fact, not going to be discharged. Supervisors who have used this tactic to rid themselves of an employee they didn't like have had to face charges of, "I was conned (intimidated, pressured) into resigning." Commissioners and board members frequently ask if the probationer was given an opportunity to resign.

The discharge or reduction in rank of the permanent employee is usually very involved. The employee normally is entitled to request an appeal hearing when the department head imposes either of these two forms of disciplinary action. An appeal hearing is not an automatic occurrence; it is generally left up to the employee to appeal the action. Obviously, the employee's offense must be in extreme contradiction to the department's standards of conduct to warrant such action, and therefore some form of investigation is indicated to establish the facts and to ensure there is adequate justification for the imposition of such a severe penalty.

The investigation usually will be conducted by an individual or unit designated by the head of the department. The investigators should be equal or superior in rank to the employee being investigated.

Building a Case

Because of the legal nature of civil service and merit system rules, supervisors must use precise methods of compiling the data for a case. Many police supervisors are reluctant to take negative disciplinary action which will require that they appear before a trial board or committee. The common complaint is that as supervisors they are often the ones on trial.

In an article on police discipline, G. Douglas Gourley stated:

> Supervisors who have this attitude are often the ones who go before boards or commissions with no evidence other than their unsupported opinions. The same supervisors would not think of going to trial in a criminal case, or permitting one of their men to do so, without gathering in advance all the available facts necessary to prove the case.

In disciplinary matters, to wait until the hearing has been scheduled to gather such evidence will be too late; it must be recorded as it takes place. A record should be made of actual incidents, events, or offenses as they occur; for modern disciplinary procedure requires the preparation of cases in a manner similar to that used in preparation of criminal cases for trial. To assist supervisors in the consistent accumulation of such evidence, proper forms should be provided.

Too often negative disciplinary action is taken as a result of repeated infractions, none of which have been discussed with the offender. Civil Service and other trial boards want not only to be presented with the details of specific infractions of the rules and the dates, times, and places that they occurred, but they also want to be assured that the supervisors have taken every reasonable opportunity to warn, reprimand and rehabilitate the offender. If supervisors have not lived up to their responsibilities in this respect, they have only themselves to blame for their embarrassment. The maintenance of records required for legal proof will result in the supervisors doing those things that too often they do not do, i.e., using warning interviews and notices of unsatisfactory conduct.[1]

Preparation

An investigation file should contain:

1. The investigative report
2. A summary of the facts
3. A biographical summary of the employee
4. A summary of the work record
5. Memorandums concerning commendations or disciplinary actions
6. A statement of recommendation by supervising officers

When the investigation is concluded and the department head has decided to reduce the rank or discharge the employee, a statement of charges and specifications must be prepared and served on the employee. This statement should designate specifically the facts for which the discharge or reduction of rank has been recommended. The method of requesting an appeal hearing and the time limitations on this right should be included. The time limit for requesting the appeal will vary, depending on the rules. When a request for appeal is received by the commission or board, a date for the hearing will be set.

Preparation for the appeal hearing is essential. In most instances the employee is entitled to have counsel representing him or her at the hearing. The department will usually be represented by a ranking officer, but generally the department's case will be presented by a member of the legal staff. The department should provide its counsel with a copy of the statement of charges and specifications. It is suggested that the complete file of the investigation and a brief similar to a prosecution summary report be given the counsel to aid in preparation of the case.

This brief would contain:

1. The offense or offenses committed by the employee

2. The date of the offense(s)

3. The location of the offense(s), if it is pertinent

4. Incriminating admissions by the employee

5. Names of witnesses and nature of their testimony

6. Previous disciplinary action taken against the employee by the department

7. Names of the investigating officers

The Appeal Hearing

The rules and methods of conducting the appeal hearing vary between jurisdictions. Legal counsel usually represents the department. There are a few common rules which should be remembered while preparing the case for the hearing:

1. Generally, in discharge or reduction of rank hearings, the burden of proof is on the department.

2. Most hearings are informal and are not conducted according to the technical rules that govern evidence and witnesses.

3. Hearsay evidence may be admitted, and all oral evidence is generally taken under either oath or affirmation.

4. In most cases, both the employee and the department are permitted to subpoena witnesses.

The defending officer or the officer's counsel usually proceeds in two ways:

1. If the charge is a *specific* violation, the defense will attempt to disprove the charge.

2. If the charge is a series of minor incidents that occurred over a period of time, the defense will attempt to prove poor supervision and persecution of the officer.

The failure to sustain the charges at an appeal hearing can cause a great deal of trouble for the department. The employee is usually much more difficult to supervise after an unsustained charge has been made against him or her. Subsequent charges must be very definite or they will be considered vindictive. The supervisor loses prestige because his or her action was not upheld and authority was successfully challenged.

The rule regarding negative disciplinary action is to take those measures which are in the best interests of the department. The preparation of the department's investigation and case should be complete in all respects, and, above all, it must be fair and direct. Preparing a case against a fellow officer is not the most appealing assignment, but it is necessary if the department is to maintain discipline. Officers assigned to this important duty should be reminded that these matters are not instigated by the department but are the result of the employee's behavior.

How to Secure Proper Discipline

The following suggestions will help supervisors develop and maintain proper discipline. These suggestions are based on the belief that it is far more important to create and maintain conditions that make corrective disciplinary actions unnecessary than it is to develop successful corrective techniques. The emphasis should be placed on preventive, not corrective, action.

Establish Written Standards

One method of securing discipline is to establish written standards of conduct. In cooperation with other supervisors in the department, a list of the standards of conduct that are expected of personnel in the department should be prepared. Each new employee should receive a copy. Everyone wants to know what is considered good behavior and proper conduct. The establishment of such standards and the understanding of them by the employees will help prevent many instances of misconduct. People have a natural sense of fairness that will cause them to abide by the rules of the game providing they know the rules.

Make Frequent Inspections

One of the more important factors in maintaining discipline is the appearance of officers and their equipment. When a supervisor insists on a smart and sharp appearance, subordinates quickly get the idea that they belong to a well-disciplined group; they will have pride in their image. This is sometimes called "spit and polish." The concept of using appearance to enhance discipline must be accepted at all supervisory levels within the organization.

The best method to achieve this "image" is frequent inspections. While inspections are not too popular at the beginning, they soon become routine and expected, and though there will be grumbling, employees will gain a sense of security and confidence in their organization. The two principal kinds of inspection are: (1) the formal, military type of inspection, when the officers are forewarned; and (2) frequent informal inspections by the supervisors.

The value of regular and frequent inspections cannot be exaggerated. Police officers learn that inspections are inevitable and that supervisors are experts at discovering those things which are not as they should be. When the rules of conduct and appearance have been enforced long enough they will be followed automatically, and the supervisor will be able to spend less time on corrective discipline. This will leave more time for the development of job enthusiasm and satisfaction which are essential to morale.

Create a Favorable Work Atmosphere and Reasonable Objectives

A favorable working atmosphere will encourage subordinates to do their best work. A good working situation involves both physical conditions and personal relationships. A word of commendation and praise for a job well done is just as essential to the maintenance of proper discipline as is the correction of a man or woman who has been guilty of some act of misconduct.

There is considerable truth in the statement that a supervisor can forget about corrective disciplinary actions if he or she sets reasonable work objectives for employees and keeps them vitally interested in reaching their objectives.

Be Open to Suggestions and Grievances

The employees should feel free to bring suggestions for improvement as well as grievances to their supervisor. Grievances do not necessarily reflect on the skill of the supervisor, but the willingness or unwillingness of subordinates to come to their supervisor with grievances—with the knowledge that he or she will be fair and open-minded in handling a problem—does. Grievances should be listened to and situations adjusted if the supervisor is in a position to take action. If not, the case should be referred to a higher authority. No employee should be denied the right to discuss a problem with a higher supervisor if the employee desires such a discussion and if the established lines of authority are followed.

Set a Good Example

The supervisor must set a good example for proper conduct. Unless the supervisor sets an example for subordinates, it is useless to expect good conduct on their part. There is often considerable truth in the old adage, "What you do speaks so loudly, I cannot hear what you say."

Be Firm and Impartial

Firm, impartial control creates respect and lessens disciplinary problems. All infractions of the standards of conduct should be corrected in private as soon as possible after they occur. In reprimanding a subordinate, the supervisor should not jump to conclusions or humiliate the individual. Instead, the supervisor should avoid general criticism and should be specific, objective, and fair.

Infractions must be corrected or employees will regard them as accepted practice. The supervisor should not allow uncorrected infractions to accumulate and become so aggravated that the first corrective act is a recommendation of separation from the department. The supervisor should call instances of misconduct or lack of self-discipline to the offender's attention and record the details of the offense. These records are helpful in the preparation and discussion of performance reports and can be referred to if formal corrective disciplinary action becomes necessary.

Avoid Overprotectiveness

A supervisor may discover over-protectiveness to be a real problem in dealing with subordinates. Police officers feel highly protective toward one another. This feeling is so strong that they will often jeopardize their own status, security, and welfare to "protect" a fellow officer. This can be understood as defensive reasoning; police officers feel that the public, as a group, does not like them, and that many people are anxious to see them penalized. Therefore, the reasoning goes, police officers must unite for their own protection. This protective feeling is not confined to the rank and file; the actions of many supervisors are also influenced by it.

This over-protectiveness, which seems to be a universal experience, must be identified and eliminated. The loyalty of police supervisors, as well as of all police officers, must be to the group or organization as a whole. An officer's primary concern must be what is best for all police officers. Certainly it is not to the group's advantage to let dereliction of duty go uncorrected, or the entire organization will be judged by the actions of a few individuals. Even at this time the over-protective attitude is weakening. As the profession gains in stature, the necessity for such feelings diminishes.

Preventive Discipline: The Clinical Approach

Finding and eliminating causes for misconduct is sometimes referred to as the clinical approach because it is based on the premise that there is a cause for every action. If all the supervisor's attention is devoted to the employee's unacceptable behavior, he or she is treating the result rather than the cause. The causes must be eliminated to prevent recurrence of the unacceptable actions.

When a subordinate has done something which is unacceptable, the supervisor should discuss the matter with this individual in private. In such discussions it is well to start with a question that will permit the person to tell his or her version of what happened. Good listening on the part of the supervisor is essential to arriving at a real understanding of what actually happened as the employee sees it. The supervisor and the employee must try to reach a mutual understanding.

The vast majority of the men and women in any department are competent, conscientious, and efficient. All organizations occasionally hire people who wilfully, thoughtlessly, or unwittingly violate the rules. Such conduct reflects unfavorably on the entire department and everyone in it. If their conduct is not corrected it will undermine the morale of other employees and lower public confidence in and respect for all police officers. Accordingly, it is to the advantage of all hard-working police employees to see that inefficient or uncooperative employees either mend their ways or become separated from the service. O. W. Wilson has stated, "Separation from the service of the inept, stupid, incompetent, or dishonest officer should not be viewed as punishment but as a device for ridding the department of persons who are not qualified to render a suitable quality of police service."[2]

Supervisors are charged with the duty of taking prompt disciplinary action when the authority has been delegated to them. If they do not have the authority to take action, they must immediately refer the matter to the proper person. The administration of prompt, fair, and effective disciplinary action is just as essential to effective operations and good employee relations as is the commendation of employees for work well done. If the supervisor takes steps to correct workers who break the rules of good behavior or to rid the department of uncooperative, incompetent, or dishonest employees, he or she will be respected by subordinates and will also increase their prestige by showing that merit is the criterion for continued employment in the police service.

Handling Personnel Complaints

The day has long since passed when a police supervisor who has knowledge of alleged misconduct can sit on a complaint, hoping that a formal complaint will not be filed with the department. To avoid intrusions such as a police review board and to fulfill responsibilities to the department, the supervisor must report the complaint and initiate a thorough and complete investigation of it to the head of the department.* This procedure must be followed even when the investigation of a serious charge thoroughly exonerates the officer. Some supervisors feel that even unjustified complaints will cloud the record of an officer. This is false reasoning; a properly written final disposition fully protects both the officer and the department.

The department must always be sure that its handling of disciplinary matters is fair, objective, and judicial. The most important internal procedure a chief of police must develop is that of internal discipline. It is essential that every complaint alleging police misconduct be immediately recorded and a copy forwarded through command channels to the chief of police. When the complaint is severe or when the press is involved, the department administrators may want to issue a statement. Since they may be deluged with calls, it is necessary that they be adequately informed when serious allegations against their personnel are being made.

The Necessity for Personnel Complaint Investigation

Personnel complaint investigations must be thorough and complete so that, whenever possible, they can be definitely settled. Delayed or incomplete investigations create an intolerable situation and are a disservice to the department, the officer involved, and the public. Good morale and discipline are at stake. Proper handling of personnel complaint investigations is essential to the maintenance of the police image for several reasons.

The first is the protection of the department. Police departments are evaluated and judged by the conduct of their individual members. An entire organization should not be subjected to public censure because of the misconduct of a few individuals.

The second is the protection of the public, which expects and demands fair and impartial law enforcement. Misconduct by officers must be detected and thoroughly investigated to insure the cooperation and continued trust of the general public. A police officer who is guilty of misconduct and is not disciplined by the department frightens people and causes them to react negatively to the department as a whole.

The third is the protection of officers. Officers must be protected against false allegations; this can be accomplished only by complete and impartial investigation. Failure to support officers who are unjustly accused will strike at the very heart of good morale.

* In some instances the matter may be investigated by an internal affairs unit of the department, but this in no way relieves the supervisor of responsibility for proper reporting. A decision must be made as to where the responsibility for investigation will be assigned. Cross jurisdiction consideration, class of violations, and name of offender are all matters for executive evaluation.

The fourth consideration is the removal of unfit personnel. Police officers who commit serious acts of misconduct or who have demonstrated that they are unfit for law enforcement work must be removed for the protection of the department, fellow officers, and the public as well as for the maintenance of morale and discipline.

The need for immediate and positive action is just as necessary for personnel complaint investigations as it is for criminal investigations because delay can result in the loss of witnesses and the loss or destruction of evidence.

Receiving the Complaint

Personnel complaints received from outside the department should be accepted and reported in writing. Complaints received from within the department should also be submitted in writing.

Anonymous complaints should be investigated within practical limits. Essentially, the same limits that govern police investigations of crime on anonymous tips should govern investigations of personnel complaints.

When an investigation discloses that an employee is suspected of a criminal offense, the person who takes the report and conducts the investigation should (through his or her immediate supervisor or, in this person's absence, through the watch commander) notify the unit or division commander as soon as possible. The unit or division commander, or the watch commander, should promptly inform the chief of police or the department's senior-officer-on-call.

Who Should Investigate the Complaint

The suspect's supervisor is usually assigned to investigate the case because he or she is directly responsible for the conduct of subordinate officers and, generally, is the one who takes the complaint. Because of a personal involvement with the individual, however, it is possible that the supervisor may be compromised. Also, the supervisor may not have the time, ability, or facilities to conduct the investigation thoroughly.

There is sometimes an advantage to assigning a different supervisor, or a personnel internal-affairs specialist to conduct the investigation. However, this system often gives line supervisors the feeling that they can shirk disciplinary situations, and very often portrays personnel investigators as the "bad guys."

Ideally, both the line supervisor and the personnel specialist could be assigned to the investigation if the conditions require this.

Evaluating the Complaint

A personnel complaint is an allegation of misconduct, and it may come from any source. The misconduct may be a criminal offense, the neglect of a duty, a violation of the department's rules, or any other conduct that reflects unfavorably on the department. The complaint should be examined with these possibilities in mind. It may actually be a personnel complaint or it may be a complaint against the department's policy, the limitations of the law, or the activities of other organizations. If the complaint is based on a misunderstanding in any of these areas, it can be cleared up by an explanation or referral to another agency.

It is imperative that the investigator of the personnel complaint keep an open mind and refrain from drawing premature conclusions, or the result may be a report that reflects the personal prejudices of the investigator. The investigator's sole allegiance must be to the department and the public. Review boards have been suggested when police conduct investigations have not been impartial.

Interviews with the Complainant and Witnesses

The initial duty of the investigation officer or supervisor is to determine the exact nature of the complaint by interviewing the complainant. The interview with the officer involved should be held as soon as possible and preferably *before* the formal interview with the complainant. Frequently, depending on the circumstances, this procedure may permit the investigator to explain the situation, mistake, or whatever, to the complainant. Caution should be used in handling complaints involving possible law violations.

In any event, no personnel investigation should be considered complete until the complainant has been personally interviewed, assuming this person is available. Investigating officers should always look for any vindictive motivation in the complainant. Such things as prior police contacts and personal misdeeds should be checked out.

Many complaints are received from drunks. The best procedure for dealing with these cases is to hold an initial interview, followed by another interview when the complainant is sober—after the hangover. The individual may be apologetic and withdraw the complaint. If the complainant holds to the facts, it is wise to check the person's background to see if there might not be some other reason for the complaint. There may have been prior complaint from the same person, or the person may have a criminal record. But *the complaint may be valid* regardless of these factors.

The location of the interview is extremely important. If it is held at the police station, it should be conducted in private, *not* at the counter! Usually, more information can be obtained if the interviews are held at the home of the complainant, or at some nonpolice facility.

The interviewing of a complainant must be preceded by an inquiry into the investigator's feelings—any fraternal feeling for a fellow officer, or prejudice that may be felt because of the complainant's personality, race, background, etc. To get the story, the investigator should then listen to the complainant's account of the incident in the complainant's own words, all the way through. The investigator should, using tact, have the citizen repeat the story several times. Each time, the complainant's emotion and anger will lessen, and the side issues will eventually disappear. The end result will be the principal issue or complaint, without the trimmings.

The investigator must not argue with or attempt to cross-examine the complainant, because the complainant may withhold information if he or she feels there will not be a fair hearing. The purpose of the interview is only to obtain the facts, not to adjudicate the complaint! However, if at the end of the interview it is obvious to the investigator that the facts do not constitute a personnel complaint as previously defined, other procedures may be taken in an attempt to close the issue. The investigator can explain the mistake in a way that will save face for the complainant, or he or she can refer the complainant to another agency or service for satisfaction, as appropriate to the case.

Investigators must refrain from irritating concerned parties—complainants, witnesses, and fellow officers. Complainants should be handled in a manner that will minimize friction. The investigator's attitude is very important. Here are some helpful guidelines for investigators to follow.

It is common for an investigator to form an opinion about the credibility of a witness at the time of the interview or during the investigation. It is appropriate and advisable to include such opinions in the report.

It is imperative to get the final written statement that will be used to substantiate the action against the officer. This statement can also provide the officer with evidence for a suit against the complainant, who can be prosecuted under any penal code sections relating to a false report of a felony or a misdemeanor to the police. The investigator should not dwell on this, because it may cause the complainant to withdraw a valid complaint. The final written statement may be taken as an affidavit which will make a perjury conviction possible.

The report of the investigation should be a clear and concise statement of all the pertinent facts of the case. The report should not include the investigator's personal opinions or conclusions as to the guilt or innocence of the accused officer or recommendations as to possible penalties.

Personnel Investigative Techniques

Investigators should apply the normal, accepted investigatory techniques, including the use of scientific aids and equipment, to the investigation of a personnel complaint. They should not forget that a timely search of a person, house, or automobile may be instrumental in resolving a complaint. The failure to conduct such a search is a disservice to the department, the concerned officers, and the complainant. These searches must be made in accordance with the law and the prevailing rules of evidence.

Employees should not be searched in public, unless it is a matter of immediate necessity. Some complaints require that an officer's personal weapons be seized for the protection of the investigators. The statements made at the time must be cautious and discrete; an explanation from the investigator can avoid bad relations should the investigation clear the employee.

If an officer refuses to cooperate in any phase of the investigation (such as answering questions or taking a test for intoxication), the officer should be given a direct order to comply. It is essential to establish that a direct order was given if a charge of insubordination is to be substantiated. If a criminal offense is involved, the suspect must be informed of his or her constitutional rights, whether the suspect is an officer or a citizen. Failure to do so will result in the loss of a conviction should the matter involve a criminal act, although this would not necessarily influence a personnel action.

Disposition of Accused Officers

The question of what should be done with the officer under investigation poses a problem. In almost all nonserious cases, the individual should be kept on duty even if assigned to special duty. If suspended from duty, the officer's morale suffers. The officer is also unavailable for contacts, may have to be reimbursed for time off, and cannot perform his or her job.

In serious cases the officer must be removed from duty while under investigation, especially if evidence indicates that the individual is unfit for service. Normally, the limits of authority are established by departmental, civil service, or merit system rules.

Accusations involving highly sensitive moral issues, such as homosexuality and child molestation, warrant taking the officer's firearm for the individual's own protection. In the early stages of investigation, officers have been known to commit suicide if actually guilty. In the event that it is necessary to detain, restrain, and confine an officer for the individual's own protection (as in the case of drunkenness, or while under the influence of drugs), it is not always necessary that the officer be formally booked. Each case must be weighed on its own merits.

Formal booking is justified in serious felony cases, and the employee's commanding officer should be the booking officer. In other felonies, the officer may not have to be booked, but can be released. But in those cases, a warrant should be obtained. The advantages of resignation may be discussed with the officer. This would not compromise a criminal complaint unless promises were made that could not be kept. A tape recording should be made of the conversation, and the officer should be made aware that the conversation is being taped.

Employee Investigation Reports

Occasionally supervisors require immediate employee reports from subordinates. Experience, however, demonstrates that it is not necessary and that it is sometimes inappropriate to require officers involved in a personnel complaint investigation to complete a routine Employee's Report at the outset. The investigating officer should question the officer concerned and relate the details of the interview in the report. This is the point at which to consider the advisability of requesting documentation from certain officers in the form of an Employee's Report.

The investigating officer should submit the report to his or her unit commander or watch commander for approval, and then forward it to the division commander or chief of police. The employee's commanding officer should take a recommendation to the chief on whether or not the complaint should be sustained.

Final Disposition by Commanding Officer

The employee's commanding officer or chief of police may ask for recommendations from supervisors after a review of the investigation. This does not relieve the commanding officer of final responsibility for disposition of the case.

An effective disposition should include the basic consideration of good morale and discipline. Practitioners of police science usually feel that derelictions should be dealt with on an individual basis. Primary consideration should be given to the motives of the officer: was the officer trying to fulfill the objectives of the department, or was this person only interested in personal desires? The individual's past disciplinary record and length of service should be considered. Suspension without pay hurts the officer's family, and this penalty should be used only in serious cases. Many officers are sufficiently punished by a written warning, admonishment, or reprimand. Officers know this will be filed in their personnel file and will be considered when they try for promotion. Criminal matters should be disposed of in accordance with the law. Any rejection of complaints should be documented in writing by the appropriate prosecutor.

When the final disposition is made, the complainant should be notified by letter of the results. This letter should be brief, without unnecessary details. If the complaint is not sustained, the complainant should be told that impartial witnesses or physical evidence, if such is the case, fail to sustain his allegation.

If the complaint is sustained, the letter should state that the officer failed to follow department procedure or policy, if such is the case, and that appropriate disciplinary action has been taken. This letter should include a statement to the effect that it is hoped that any future contacts with members of the department will be more favorable.

DISCUSSION QUESTIONS

1. What are the negative incentives or punitive disciplinary actions most frequently used in police agencies? In what circumstances might each be used?
2. Discuss the positive and negative connotations of discipline. How are they related to training?
3. What are the results of poor discipline? Good discipline?
4. What is the role of the supervisor in relation to discipline? How should responsibilities be discharged?
5. What are the points to remember when contemplating corrective disciplinary action?

NOTES

1. G. Douglas Gourley, "Police Discipline," *Journal of Criminal Law and Criminology,* Vol. 41, No. 1 (May-June, 1950).
2. O. W. Wilson, *Police Administration* (New York: McGraw-Hill, 1963), p. 173.

11 Communication

Communication, reduced to its essentials, can be defined as the transfer of intelligence or feeling from one being to another. This simple definition covers everything from a friendly nod of greeting to the most complicated code transmitted via satellite.[1] It should be noticed that there are two parts or necessary conditions embodied in the definition. The first is that there must be intelligence or feeling; the most eloquent speaker imaginable would not be communicating if he or she simply read words at random from a dictionary. On the other hand, a dog communicates quite well when it bares its teeth and growls! The other necessary element, the transfer, is equally obvious; if a person is speaking a language that no one understands, there is no communication.

Everyone communicates every day, but some people are much better at it than others. These people are communicating *effectively*. For effective communication one more element must be added to the definition: the information transferred must be undistorted. The management of a police department might formulate a highly workable tactical plan for controlling an insurrection, with alternatives to cover many different contingencies, but if the supervisor simply tells subordinates, "Get over there and take care of that mob!" the communication has not been very effective. It has been distorted during the transfer from management to patrol officers.

Communication is the binding material of any organization. It enables the group to function as a unit, rather than as a number of individuals. The larger the group, the more difficult it is to ensure effective communication among all its members. Even more important, the larger the group, the more vital are its communications.

COMMUNICATION

A two-person police force in a small village could conceivably get along with no communications; each officer could roam the village and deal with problems as they arose. But imagine the havoc and confusion if the thousands of officers of the New York police force each decided to operate as an independent agent, neither receiving direction nor cooperating with one another. Because communications increase in both difficulty and importance as the organization grows larger, systems of communications are developed, such as the positional communication system which will be discussed in the next section.

In the study of any subject, it is useful to categorize so that one section may be studied at a time. Communications could be divided in many different ways: by technique involved, by type of person doing the communicating, by subject matter, etc. Since this discussion relates to police supervisors, it was decided to break the subject into three areas. The first is communication to produce action (orders, directives). The second is communication to inform (reports, memos, conversations). And the third is communication to elicit information and possibly produce a change in a person (the interview). This last subject will be treated in a separate chapter. Many of the techniques and considerations are the same for the first two categories so they will often be treated together, but the reader should be alert and watch for differences.

Positional Communication

Positional communication is a system which works well in organizations that are structured on a pyramid basis. Its primary function is control of information flow and directives. There are three basic problems that a positional communication system solves: it ensures that everyone who should be informed is informed; it restricts the dissemination of information to those who should have access to it; and it filters information so that an executive does not have to deal with a problem that a subordinate would be able to handle.

Vertical Communication

For the system to work, everyone must understand it and follow its rules. Within the pyramid, information may proceed upward and inward, downward and outward, and laterally. The originator of information which is to move downward must do two things: stipulate the level to which the information is to go, and make sure that every person on the level immediately below receives it. Those people are responsible for passing the information to those immediately below them; thus, the information proceeds downwards, level by level, until it reaches that level specified by the originator. Several things should be noticed: no person will have to supply information to more than seven or eight people; everyone between the originator and the specified level will be informed; and no one below the specified level will have access to the information.

The originator of information that is to move upwards has a simple task: to give the information to the immediate superior and to no one else. That person will either deal with the problem or give the information to his or her immediate superior. When the information gets high enough for someone to act upon it, it will stop. No one at a higher level will have to take time to evaluate it.

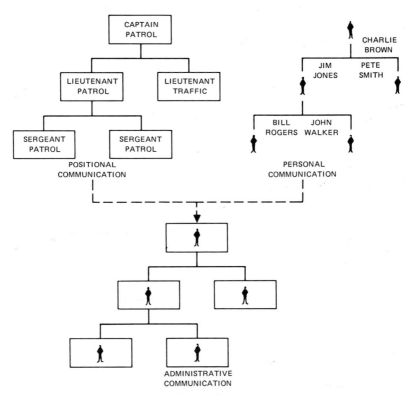

FIGURE 7. The Positional Communication System

Horizontal or Lateral Communication

Information that is to move laterally may be handled in two ways. If the number of people to be informed is small, the originator can simply do it personally; if the number is large (if, for instance, one supervisor wishes to communicate with all other supervisors), the information can be directed upwards until it reaches the level which covers all those for whom the information is intended. The person at the top then directs the information downwards in the normal fashion. When the originator receives the information back, the person knows that everyone on his or her level has also received it.

Some of the most persistent and acute problems of administration, particularly in large organizations, stem from deficiencies in lateral communication. One of the principal responsibilities of many staff people in large agencies is to transmit information to positions and units at the same level. In the police agency, lateral communication is extremely important in coordinating the activities of the various units and divisions of the department and ensuring uniform interpretation of policy.

The Human Element

The only real flaw in the positional communication system is that the positions are filled by human beings who have a natural tendency to short-circuit the system. A chief decides to mimeograph an order and pass it out to the employees—and three supervisors don't find out about the new order until their subordinates tell them about it. Or a patrol officer can't find a supervisor and reports directly to a lieutenant who is too busy to deal with the problem. The lieutenant isn't worried because he is sure the supervisor will take care of it—but the supervisor is never made aware of the problem. It is obvious that it takes only a few irresponsible people in an organization to make a shambles of the communications network.

The three basic rules to be remembered are:

> 1. When communication is upwards, *only* the immediate superior should be informed.
>
> 2. When communication is downwards, *everyone* on the level immediately below the originator must be informed.
>
> 3. When communication is lateral, it can move upwards or downwards, but no decision should be made before each situation is evaluated. The staff should be used whenever possible.

Principles of Effective Communication

Clarity

The first prerequisite for effective communication is clarity. In speaking or writing, if the presentation is not clear it will not be understood. There are times when a lack of clarity only results in embarrassment or provides some light comedy. For example, a meeting is scheduled to take place on the corner of Third and Main. But the communicator neglects to mention that the meeting is to take place in New York, and the recipient of the incomplete message waits for hours in New Haven. A similar misunderstanding during a bank robbery, however, would not be comic; it could even cost someone's life.

It is a real paradox that clarity, which first and foremost requires simplicity, is seldom achieved because people find it easier to be complex! Too many words and too little information are the worst enemies of clarity. Unclear messages are caused by poor organization, the wrong choice of word, and poor grammar.

What can be done? First, organization of the original information can be improved. Before the message is written or spoken, it should be dissected. The five key words of the journalist—who, what, when, where, and why—are a big help here. Jotting down the response to those words and using the result as a basis for the message would provide more organization than is seen in 90 percent of the communications that take place in any large corporation. When longer messages or reports are to be written, an outline will help immensely.

The improvement of diction is the next step to be taken in the achievement of clarity. Diction concerns the correct choice of words, as well as pronunciation and enunciation. There are two common faults to overcome: the tendency to be careless in choosing words and the tendency to use long words and roundabout phrases where simpler language would be adequate. The former tendency reveals itself in sentences such as, "Why don't you investigate . . . ?" when what is meant is a direct order. As quoted, this phrase is a suggestion, not a directive. Often a general word, such as "gun," will be used when a more specific term, such as "semi-automatic rifle," is required. Sometimes, of course, the wrong word is used because the distinction between two words is not clearly understood. For instance, a man might be charged with burglary when he is actually guilty of theft. The second tendency, to use long words and cumbersome phrases, results in monstrosities such as, "It is desirable that you implement this directive at your earliest possible convenience." This could probably be translated, "Do this right away." Although the use of some jargon is inevitable, it should be avoided whenever possible. The continual use of jargon very often marks the user as insecure and incompetent. The rule should be: Use painstaking care that what is said corresponds with what is intended. A word should never be used unless both parties know exactly what is meant and there is no simpler word that will do the job. One of the best-known sentences of modern times is, "The only thing we have to fear is fear itself." The longest word has six letters.

The only real answer to the problem of poor grammar is study. Many people do not even know what grammar is, let alone the difference between correct and incorrect grammar. The English language uses two devices to associate concepts: the function word (prepositions, articles, etc.) and word order, or the relations of words in the sentence. The study of these two devices constitutes grammar. Extensive reading helps develop an intuitive feel for it, but there is no substitute for a few hours of hard work with a student's textbook. Failing that, the best solution, again, is to keep things simple. Short, declarative sentences in the present tense will seldom be misunderstood.

Here are a few practical tips. The use of a comparison often helps to clarify. If you say, "There were one hundred burglaries this year," it doesn't mean very much, but if you compare it to last year, when there were five hundred, it means quite a bit more. Sometimes there are established standards that can be used for comparisons.

When writing instructions, it is best to try to imagine every single step that must be taken and then to write them down in just that order. The most significant reason for failure to carry out orders is misunderstanding the instructions. Although every necessary step must be clearly stated, too much detail should be avoided; people tend not to read boring material. One of the most common mistakes that supervisors make is to assume that an employee knows something that, in fact, he or she does not. The best policy is not to assume anything. It is very hard for a supervisor to go to an officer's widow and tell her, "I *assumed* he knew the suspect was armed and dangerous."

Consistency and Continuity

Consistency and continuity are two closely related, extremely important aspects of any communications system. The literal meanings of the words are quite obvious, but the implications to the police service are often not obvious at all. Consistency demands that orders, directives, policies and the like, reflect the same point of view, whether being transmitted by different people or at different times.

Continuity demands that orders be enforced, that policies be adhered to, that programs be followed up, and so on. An example will make the distinction clear. If the employees of one unit were encouraged to submit suggestions for bettering the operation, while the employees of a second unit were told not to, this would reveal an inconsistency. If, on the other hand, employees were asked to submit suggestions which were never evaluated, there would be a lack of continuity.

Failure in either of these areas signifies a breakdown of the communications apparatus. Policies and directives reflect management's interest in some particular subject area. If they are inconsistent, it means either that the management is uncoordinated or that some lower level is failing to communicate, or transmit, management's interests to the employees. If there is a lack of continuity, either the policies and directives were not designed properly—with provisions for follow-up—or someone at a lower level, whose responsibility it was to follow up, has failed to do so.

An exception to a rule or policy constitutes an inconsistency, although not the same type of inconsistency described above. If all personnel are required to report for work at 8 A.M. except John Jones, who is allowed for some reason to report at 8:30, an exception to the rule, or an inconsistency, has been created. There may be very good reasons for this exception, and, indeed, the rule might even provide for some exceptions. But this does not make the exception desirable. Exceptions have a tendency to establish precedents that over a period of time accumulate and establish new rules. The rules thus created are haphazard, based, as they are, on expediency and circumstances. They are not the result of logical planning. If the rules and policies are allowed to evolve in this fashion, the communications system of the organization will soon resemble the street plans of our major cities, which have a similar historical evolution!

It is a common practice of administrators and supervisors to write exceptions into their rules and policies for their own protection. By using jargon, abstruse terms and bureaucratic gobbledygook, "hidden outs" can be included that will not be detected—if they are read at all. These will allow the initiator to claim all the credit if things go well—and have an alibi if things go wrong! Needless to say, this practice runs contrary to all principles of good communication and will expose the perpetrator as a poor leader who is unwilling to accept responsibility.

The practice discussed above gave rise to a technique called the "exception principle," which is quite useful. According to this principle, reports and orders should be read, not for the routine and satisfactory aspects, but for the exceptions. Here will be found the clues to what is really happening. For example, if a report is submitted describing a project to be initiated and, in general, it looks worthwhile and well thought out, the exceptions are what should be looked for. Perhaps all the requirements for each phase of the project have been listed, except that there is no cost estimate for "Phase Three." After approving the project and spending enough time and money on it so that there is a virtually irrevocable commitment, the superior may find that the cost of "Phase Three" is more than the total cost of the rest of the project. And further, the supervisor may discover that the initiator of the project knew the phase would be very expensive and didn't include that factor because he or she also knew the project would not be approved if cost was included. Following the exception principle, the superior would have zeroed in on that one particular item and asked, "What is 'Phase Three' going to cost?" thereby saving a great deal of later embarrassment.

Although a great deal has been said against them, it should not be thought that there is no place for exceptions and inconsistencies. There must always be some flexibility, and the more general a rule or policy is, the more flexible it must be. To establish a rule for a 1,000-person organization that states, "Everyone will be at work at 8 A.M." and allow no exceptions would be absurd. The principle is "No Exceptions," but the principle must be tempered with judgment. The student should have noted that this paragraph represents an exception to the principal that is being discussed.

Adequacy

Each person in a communication system should receive all the information needed to function and no more than that. If a person receives too little information, he or she will be forced to make unsound judgments and decisions; if the individual receives too much, time will be spent inefficiently reading unnecessary reports and memos. All reports and other communications should be evaluated and filtered before being passed on, either up or down the line.

Timing

The timing of a communication can be very important from a practical as well as a psychological viewpoint. The Army is famous for such errors in timing as issuing overcoats to its soldiers in the summer and supplying them with mosquito repellent in the winter. Errors less gross, however, are committed every day. To order personnel to attend a special training session immediately after they have been working overtime and are very tired, would be one example of poor timing.

Timeliness

Out-of-date information is as worthless as none at all—or worse. If a report or order is delayed, it ceases to be a control-medium and becomes a historical document. The responsibility of the supervisor does not end with issuing an order or writing a report. It is essential that he or she review and revise instructions and reporting systems periodically to eliminate or forestall obsolescence.

Distribution

The most carefully prepared message may be undercut in its transmission. The communicator must ask, "What audience do I wish to reach?" A police agency has many different audiences, a few of which are: certain positions (all traffic sergeants), certain units (the entire traffic division), a certain designated individual (watch commander of traffic, third platoon), etc.

The audience must be designated before the message is transmitted. *Who* is to be told is often just as important as *what* is to be told, because the question of *who* often determines *how* the telling is to be done. The person receiving a message must always know its purpose because the individual has the authority to act upon the message or the responsibility for action, and because the individual may profit from the information. It is important that there *be* a reason. Nothing is quite so frustrating as receiving a message and not having the vaguest idea why it was sent.

Three Categories of Information

There are three categories of information to be communicated to subordinates. The first is information that should be communicated *immediately*. Such information directly concerns an employee or his or her job—work assignments, methods of operations, rules and regulations, duties and responsibilities, assessments of performance, etc.

The second category is information that should be communicated *as soon as possible*. This information concerns knowledge and attitudes necessary to coordinate each employee's work with that of other people or units. Such information usually pertains to the future. Examples are: policy matters; department organization; pay and fringe benefits; the place of the individual job in the overall scheme of things; expected standards of personal conduct; work expected; and anticipated changes in operation procedures, systems, or personnel that may influence the individual, the job, or the unit at some future date.

The last category contains information that may or may not be communicated. This information deals with areas such as future plans for growth and changes in organization, department policy, and other general matters.

Information that is necessary for people at one level may be less essential for people at another level. There are no hard-and-fast rules. Good judgment and knowledge of people and what they want to know is helpful. Employees have an almost unlimited desire for information. Surveys indicate that few employees think that they receive too much information. It is better, when making a distribution list, to include someone who does not need the information than to overlook someone who does. People who are forgotten will become resentful.

The supervisor must distinguish between what is needed and what is wanted in communicating with subordinates. The supervisor must give them information that will make them realize that they are important to the department, and that will satisfy their needs for attention and status. In addition, the supervisor must communicate that which is necessary to arouse feelings of opportunity and of security. When subordinates are well informed, the anxiety and aimless questions that cause confusion and indifference are eliminated.

Subordinates should be told everything they might eventually learn for themselves. In this way, the facts can be given truthfully and constructively, distortion through lack of information can be prevented, and the supervisor can establish a reputation for being an open and primary source of trustworthy information.

Uniformity

One of the most difficult problems confronting police management is the achievement of a proper balance between flexibility and uniformity in administration and operations. Uniformity is not entirely separable from the principle of consistency discussed earlier. The smooth operation of the department depends to a great degree on uniformity. Standard operating procedures and uniform systems of orders and reports have demonstrated their value in police operations. However, at times, uniformity must bow to flexibility because of the different situations that arise and the individuals who are involved. For example, the use of standard forms for various types of investigations may limit the initiative of an officer who is conducting an investigation. The officer may feel that all responsibility has been properly discharged when all the blanks on the form have been filled in. An investigative report form should be designed to place a floor under an officer's efforts rather than a ceiling over them. Some provision must be made for adaptation, and supervisors should encourage initiative on the part of subordinates.

Interest and Acceptance

The purpose of most communication is the prompting of some kind of action or change in attitude. If a directive is issued, it is expected that some change in the operation will take place; if a rating interview is conducted, a change in attitude toward some phase of the job is desired. But there will be no action or change unless the recipient of the communication accepts it. And there will very likely be no acceptance unless the individual is interested in what is being communicated. So interest and acceptance go hand-in-hand and both are vitally necessary to effective communication.

The first question is: how is interest generated? It has become increasingly clear in recent years that one of the poorest ways to create interest in a program is to deliver a lecture. Unless the lecturer is very talented (and most amateurs aren't), the audience will not even become aware of the lecturer's ideas, let alone interested enough to translate them into action. Since one-to-one communication is extremely inefficient when communication among many people is desired, experiments have been conducted in group discussion.

Group Discussion

In group discussion, a problem is presented along with its complications and some possible solutions. The group then evaluates the problem, discusses the given solutions and puts forth any other solutions that it can produce. It also discusses ways of implementing the solutions. The theory is that when a group chooses a solution in this manner and agrees upon a course of action, it will regard the problem as its own, be interested in it, and carry out the action decided upon. Following are some summaries of studies that relate to this problem.

In the classic "Studies in Group Decision,"[2] Kurt Lewin found that 30 percent more people followed a recommended course of action when they decided mutually, in group discussion, the action that would be taken, than did a like group of people who were merely told in a lecture what they should do.

Levine and Butler[3] conducted a study of methods of changing socially undesirable behavior. The study's objective was to persuade a group of supervisors to rate workers under them more objectively. Again, after the supervisors had been lectured to, there was no change in their method of rating. However, when the supervisors had an opportunity to discuss the problem with their superior and to decide for themselves, after having been given the facts, they actually became substantially more objective in their ratings. The principal conclusion of this study was that "once a *group* arrives at a decision to act, the members take on that decision as their own and act favorably upon it."

In another study pertaining to the attitudes of people toward various issues and problems, Karl R. Robinson drew similar conclusions.[4] He found that changes take place in individuals' attitudes towards problems as a result of their participation in group discussion and that argumentative and dogmatic persons undergo slightly larger changes of opinion than do those who are cooperative and friendly. He also found that 93 percent of those who engaged in discussion of the problems had acquired additional information pertaining either to the background or to the possible solutions of the problems. Finally, from reports of the observers, it was learned that three-fourths of the group engaging in discussion were able to agree completely on solutions to the problems discussed.

In another study, Goodwin B. Watson[5] compared the intellectual efficiency of a group to the efficiency of the individuals who were working by themselves. He discovered that groups working cooperatively were better able to accomplish some tasks than were individuals working alone. Furthermore, he found that when the groups were enlarged, the assigned tasks were accomplished even more efficiently.

Selling Orders to Subordinates

Selling orders is important in gaining willing cooperation. It establishes a positive attitude, creates the will to work, gains willing, enthusiastic effort, and secures intelligent cooperation.

To sell an order, the supervisor must gain approval for it by showing that it is important, by demonstrating its soundness, and by showing that it is practical. If the supervisor is assigning work he or she should make an effort to see that subordinates understand the reason for the assignment and, most importantly, that they know exactly what they are to do. This subject will be thoroughly covered in Chapter 15.

One of the most important factors in selling orders to subordinates is the manner in which the supervisor conveys the orders. A negative attitude will give the subordinate the general impression that "Here's another one of those things the top brass dreamed up when they had nothing better to do with their time." One could hardly expect an enthusiastic response to the order from the subordinate.

In selling orders it is also important to provide a personal reason for compliance if possible. For example: Which of the following orders is likely to be best received and has the best chance of being obeyed? (Note that in each case the order is also integrated with other operations.)

> "Don't forget to turn on the outside speaker when you leave the car to talk to a violator. The radio dispatchers are having to call some of you too long before getting an answer." Or, "We are working one-person cars night and day. There are going to be times when you or another officer will need help in a hurry. Every officer on duty must be prepared to answer such a call at once, without any lost time whatsoever. The department has made it easy for you to hear your calls when you are out of the car by installing outside speakers. But, the outside speakers won't work unless you turn them on before leaving the car. Please make it a point to turn on the outside speaker *every time* you get out of the car, if you are going to be within hearing distance. And if you are going to be away from the car, tell the dispatcher where, why, and for approximately how long. If you do these things it will make it possible for you to give emergency help to your fellow officers, and for them to be ready to give you help in a hurry."

It is almost always possible to find some aspect of the instruction which will make it personally important to the individual.

Good leadership will simplify the task of giving orders. A supervisor who has the respect, confidence, and loyalty of subordinates will find that he or she has little difficulty in getting instructions and policies carried out. But a person cannot rely on high morale and good attitudes alone.

Men and women fail to carry out orders properly becasue they do not know what to do, they do not know how to do it, or they do not want to do it. The supervisor's job is to see to it that subordinates know what to do and how to do it, and that they do it properly.

An order should be worded courteously. Whether written or verbal, the phrasing of an order will have a good deal of influence on its reception. Poor wording can arouse resentment and anger even though the order is proper and legitimate. Terms which are easily accepted, such as "Would you," "Please," and "I will appreciate" should be used, and peremptory phrases which build barriers, such as "Go and do that," "I want you to," and "Get out" should be avoided.

The supervisor should remember the psychology of human differences. People react differently to words, phrases, and tasks. Some individuals will catch on quickly, others slowly. Some will anticipate, think they understand before they really do, and not pay attention to details. Instructions must be tailor-made to fit the individual.

Before giving an order, no matter how minor it is, the supervisor should determine if it is important and worth doing. Also, the supervisor should ascertain if it is practical, if it will work, and if it will take more time than it is worth. Answering the following questions beforehand will make the final product (the order) much easier to sell. Can the job be carried out? Is it both possible and reasonable? Is there any conflict with other orders, policies, or practices, written or unwritten? Will it require excessive instruction or training time? Will it require excessive follow-up to insure compliance? Can subordinates have a voice in preparing the order, and if so, how? If difficulty can be expected and the plan is not really important, the supervisor should not enact it.

Follow-up

Follow-up on orders is an essential part of supervisory control. The individual who gives an order must see to it that it has been properly executed. If there is no follow-up, supervisory time and effort are wasted, the time subordinates spend receiving and considering the order is wasted, and subordinates are confused and uncertain while the order is theoretically in effect but not being enforced. Also, the supervisor loses prestige because of a failure to secure compliance with his or her instructions, and it becomes more difficult for this person to have future orders accepted.

Without follow-up, the supervisor will never know whether instructions have really been sold to subordinates. Some subordinates may only agree that instructions are proper and workable because they have found that the supervisor will make it uncomfortable for them if they disagree. Agreement does not necessarily mean cooperative and prompt compliance. Compliance may be slow, or there may be passive resistance.

Subordinates may agree with the instructions at first, but object later when they have had a chance to think about it. They may comply at first but slack off later, particularly when they note that there is no follow-up by the supervisor.

A follow-up reminder demonstrates the supervisor's interest in the matter. It also serves to reemphasize the supervisor's belief that the orders are sound and important, and provides the subordinate with a second opportunity for compliance.

Follow-up by inspections and reports may be accomplished in several ways. The supervisor may check the standard reports and records, which usually contain the desired information. If really needed, special reports may be utilized. But the supervisor should remember that this adds to his or her own workload and to the workload of subordinates. The supervisor should be sure to discontinue special reports as soon as they have fulfilled their purpose.

The supervisor may make inspections at the level of immediate supervision. Every opportunity should be taken to observe the operation and examine the work being done to ensure its compliance with orders.

The frequency of follow-up reports and inspections will vary. Some follow-up checks accomplish little because the interval between them is too long. After instructions are given, checks should begin as soon as partial compliance is accomplished, and should be repeated at sufficiently close intervals to keep the orders fresh in the minds of subordinates.

Barriers to Effective Communication

In addition to learning the techniques of good communication, most people find that they must also break down several barriers before they can become effective communicators. Chief among these is the inability or unwillingness to listen. Communication is a two-way street; information must be taken in, evaluated, filtered, and organized before it can be communicated to someone else. One of the most important ways of taking in information is by listening.

Another barrier to communication is the insulation layer. Whenever information is relayed through several people, some distortion occurs. It may take the form of a word substitution that subtly changes the meaning, it might be an important phrase or qualifier that is left out, or it might be just a change of tone or emphasis. Gossip is a good example of this in everyday life. For example, the visual action is a warm greeting. It is observed and reported, and the story travels from person to person throughout the office. Finally it reaches the employee's spouse as a kiss or an overture. After being transmitted by twenty-five people or so, an innocent greeting can become grounds for divorce. And each person who transmitted the information thought the incident was being reported accurately. People who are quoted frequently in the newspapers complain that what appears in print is not what they said at all. Yet no one in the chain of people who handled the information consciously distorted it.

The third barrier to communication concerns the question of written versus verbal communication. Written communication suffers less distortion as it is transmitted, whereas verbal communication more accurately reflects the communicator's thoughts. Which should be used in various circumstances? These three barriers to effective communication are explored in detail in this section.

Learning to Listen

In a study first conducted by Dr. Paul Rankin,[6] and later repeated by several independent researchers, the following interesting data were obtained. First, it was found that 70 percent of an individual's waking day was spent communicating. This of course is an averaged figure; some people spend more time communicating than others. Of that time, it was found that 45 percent was spent listening, 30 percent speaking, 16 percent reading, and only 9 percent writing. Logically, then, it would seem as though much more time should be spent learning how to listen than learning how to write! In fact, of all the methods of communication mentioned, listening is the only one which is almost never studied formally at all. It is assumed that people learn to listen "naturally," whatever that means.

Good listening, like good writing, is an art. As such, it requires a knowledge of its principles and constant practice to achieve proficiency. Any conversation can furnish an opportunity to practice listening. The trick is to become *consciously* aware of listening, to apply the principles, and then to evaluate afterwards. How much of what the speaker said is really remembered? Here are the principles as developed by Nichols and Lewis.[7]

1. *Expose yourself to difficult listening material.* It is hard to improve your listening habits if you watch nothing more challenging than Westerns.

2. *Be interested in all topics.* Why? It is the best way to gain new information. It is the best way to grow culturally. It is the best way and surest route to social maturity.

3. *Adjust to the speaker.* Not all speakers are good speakers. Ignore those things about a speaker's delivery that distract you and concentrate on what is being said. This is the time to actively pick the speaker's brain for what you can get out of it.

4. *Listening is hard work.* One must work at listening. You can't allow your own personal problems, frustrations, fears, etc. to get in the way of your listening. There is a time for these, but not while you are listening.

5. *Put yourself in a good position to see and hear.* There seems to be some magic fascination about the back row in our society; it seems

that's the place most people in an audience like to go. However, it's bad for listening.

6. *Avoid overreacting to "loaded" words.* "Jew," "chiseler," "Red," "farmer," "nigger," "yokel," "cop," and "welsher" are a few examples. If you allow yourself to react to loaded words, then you may miss the point that the speaker is trying to make.

7. *Avoid overreacting to emotion-rousing points.* Many times, some of the points that a speaker brings up may provoke very strong feelings. As the old saying goes, you may "see red." Allow yourself to react this way, and you may miss the whole idea of what the speaker is trying to say.

8. *Recognize central ideas.* When listening to long presentations, don't try to get all the details but, rather, listen for central ideas. A good question to be asking yourself is: What is the speaker really driving at? What's the main idea here?

9. *Think while you listen.* It is a proven fact that we can think four, five, or more times faster than any speaker can talk! While you are listening, review what the speaker has already said. Try to think ahead to what the speaker is going to say. Evaluate what the person has said. Look for the ways in which he or she is supporting arguments.

10. *Use notes when necessary.* But be careful when doing so. Don't ever try to write down everything you hear—you can't. Here again, take down only the main ideas, the central ideas being presented. Later they will serve to remind you of the details.

Layers of Insulation

The longer the chain of command in an organization becomes, according to G. Douglas Gourley,[8] the more difficult communication becomes in both directions. This is because a "layer of insulation" accompanies each link in the chain. More simply, each person who has a chance to handle the information also has a chance to delay it, to distort it, to "edit" it, to add something to it, or simply not to transmit it at all. And all of these things happen to information as it journeys up and down the chain of command. Here are some of the reasons.

The tendency to procrastinate, or simple laziness, accounts for much of the "insulation" effect. Each person receives the information, holds it for a day or two until dealing with it, and by the time it gets down to the ranks, it's old news. Whatever luster or urgency it had has been long since lost. One method of dealing with this problem is to mark the date and time on each piece of information as it arrives and as it leaves. If this becomes a standard procedure, those who are holding up the information are spotlighted. Laziness will usually evaporate in the heat of a spotlight.

There is a natural human tendency to want to broadcast good news and hide bad news. This is especially true when the news concerns the person who is supposed to be transmitting it. If a supervisor has a report that tends to show him or her in a bad light, the individual may simply "forget" to pass it on. This practice of filtering and suppressing information must be strongly discouraged. If an executive is to formulate sound policy, this person must have accurate information. If the executive hears only about the work that is done well, any policies made will be necessarily inadequate. The solution to this problem is to not tolerate it. The first offense should merit strong censure.

The most difficult "insulation" to deal with is the one based on a psychological difficulty. To start with, it is not always easy to locate the trouble. Each link in the chain of command will fail to pass on some information for perfectly just and valid reasons. For instance, a supervisor will not pass on a report concerning a subordinate if it is a minor disciplinary matter, preferring to handle it personally rather than waste the time of someone higher up. This was discussed earlier in the chapter. Each person must make decisions about which information must be transmitted and which must not be. If a person is failing to transmit or transmitting late for the reasons already discussed, it quickly becomes obvious. If the difficulty is psychological, though, there won't be any pattern that is easily discernible.

Some supervisors and executives like to think of themselves as "masters of mystery," possessing knowledge that no one else has. Generally these are little people who are satisfying a secret desire for power. It is a characteristic of the immature. For others, this becomes a defense mechanism to allay feelings of insecurity. If the person is occupying a position for which he or she has not the ability or the skill, this mediocrity is hidden under a blanket of mystery. The theory here is that if no one knows what the individual is doing, no one is in a position to say whether or not it is being done well or poorly. Still another type is the person who likes to manipulate. This person's contributions to misinformation can be very erratic. The individual might decide that a chief's plan is poorly conceived and delay the information about it until it is too late to put the plan into operation. Or the individual might think the number of officers specified for an operation is inadequate and arrange a "mistake" in retyping the report. Where the number was fifty, it will now be one hundred. Since this person is, in one sense, playing a game we might call "The Master Manipulator," it will be as important to that individual to avoid being caught as it was to arrange the manipulation. The efforts of this person will usually be well covered.

None of these activities will be easily discovered and, even when discovered, there might be little that can be done. Sometimes simply explaining to such a person the reasons for his or her actions will make some improvement. In any event, punitive discipline will seldom have a beneficial effect.

The term, "layers of insulation," has been used throughout this discussion, but there is a more fundamental term; it is distortion. All information that is transmitted, in whatever medium, is distorted, either in time or in quality. Thoughts that are verbalized do not come out quite as the person intended, just as a record, each time it is played, picks up a little noise, and a radio signal picks up static each time it is relayed. In deciding how a report is to be delivered—whether it is to be simply dropped in the in-basket or handed to the person—a supervisor adds distortion. The importance of the message can also be added to or subtracted from by this person. If he or she chooses to say, "This is from the chief," still more is added. Even a secretary distorts when selecting the quality of paper to be used to type a directive, or when inadvertently misplacing a comma.

There is no possible way to eliminate all distortion in the transmission of information, but distortion can be minimized with care and objectivity. Each person who is charged with relaying information must make an effort to understand exactly the originator's intention. The individual must then try to carry out that intention as best he or she can, fully aware that the effort will be imperfect, but determined to keep the imperfection to a minimum.

Written or Verbal Orders

Making a choice between written or verbal communication amounts to making a choice between types of distortion. No written communication can have the degree of subtlety that is present in even the simplest verbal order. For instance, a superior tells a supervisor to have John or Mary do a particular task. By simply raising an eyebrow when saying the person's name, the supervisor can convey the idea that this choice is a bit dissatisfying, or that John or Mary should be carefully watched on this job, or any of a myriad of other things that only that particular supervisor will be able to interpret. None of this could be conveyed in a written order.

On the other hand, to try to communicate verbally the traffic control plan for a large section of the city would be nonsense. In this case, nuance and subtle implication are unimportant; what matters is the communication of a mass of details. Also, the same communication must be made to many people, and only through written transmissions can minimum distortion be obtained. Even when something relatively simple is to be communicated to many people, the written communication is preferred. None of those who receive the message will have the benefit of various possible shades of meaning, but all will get a very close approximation of the same message. But when great tact, diplomacy, and empathy are required, there is no substitute for direct, face-to-face communication. The following are suggestions for specific situations:

> The use of verbal orders is indicated when the job is not complex and confusion or misunderstanding is not probable, or when the job has

been done before and has become standardized. Verbal orders are used in directing assignments by radio, in the handling of emergencies, and when time is an important element.

Written orders should be used when orders are standing, when they are nonroutine, when they are complex enough that misunderstanding is reasonably possible, when several persons are involved and all must have the same understanding, when coordination is necessary, and when control and follow-up will be simplified.

The advantages of written orders are: they force the supervisor to organize instructions systematically; there is less possibility of overlooking an important item; and all persons receive the same instructions and are more likely to have the same understanding. Written orders permit the recipient to refresh his or her memory by referring to them for details, and they serve as a reminder if the activity is not to be done immediately. File copies will facilitate follow-up. They serve as evidence of the precise instructions given in the event of noncompliance, and they prevent lost time and effort because of misunderstanding. Whenever a subordinate performs a job improperly because he or she does not understand it, the time required to clear up the misunderstanding is usually far greater than the time required to make the instructions clear in the first place.

DISCUSSION QUESTIONS

1. What is the importance of communication to management?
2. What are the essential elements of an act of communication? What is the significance of each?
3. What are the different types of communication and what is the significance of each?
4. What is the meaning and importance of positional communication?
5. What are some of the methods of communication other than verbal and written?
6. In what directions do communications flow within the organization? What sorts of items flow in each of the channels?
7. What are the basic principles of effective communication? How is each of them important in supervision?
8. What are some of the problems of communication in management? How are they overcome?
9. What are the essentials in giving orders? Why are they important?
10. How should the supervisor follow up orders?

NOTES

1. Many of the matters discussed in this chapter are developed at length in Charles Redfield, *Communications in Management* (Chicago: University of Chicago Press, 1955). See especially pp. 11-12, and 26-41.
2. Kurt Lewin, "Studies in Group Decision" in Dorwin Cartwright and Alvin Zander, *Group Dynamics* (Row Peterson and Co., 1953), pp. 287-301.
3. Jacob Levine and John Butler, "Lecture vs. Group Decision in Changing Behavior," in *Group Dynamics* (Row Peterson and Co., 1953), pp. 280-286.
4. Karl F. Robinson, "An Experimental Study of the Effects of Group Discussion Upon the Social Attitudes of College Students," *Speech Monographs* (1941), pp. 34-57.
5. Goodwin B. Watson, "Do Groups Think More Efficiently Than Individuals," *Journal of Abnormal and Social Psychology* (1938), pp. 328-336.
6. Paul Rankin, "The Importance of Listening Ability," *English Journal* Col. Ed. (October, 1928), pp. 623-630.
7. Ralph G. Nichols and Thomas R. Lewis, *Listening and Speaking* (William C. Brown and Co., 1954).
8. G. Douglas Gourley, "Encouraging Compliance With Policies," *Police* (November-December, 1960), p. 55.

12 Counseling and Interviewing

It has been said that if police supervisors devoted the same time and skill to interviewing and counseling employees as they do to interviewing witnesses and suspects, the morale of many police organizations would be greatly improved. Regardless of the truth of this statement, there is no doubt that morale in the police service would be improved by the friendliness and mutual confidence that results from good interviewing and counseling techniques. Mutual respect would certainly be achieved if all supervisors and other employees practiced good human relations.

Be a Good Listener

An interview is an exchange of ideas between two or more people, and the supervisor should realize that at least two attitudes will be expressed—the supervisor's and the interviewee's.

A large percentage of a supervisor's time is spent talking to people. The higher his or her rank, the more the individual may talk, and the more likely that person is to be involved in major matters that require highly developed communication skills. At any rate, the old expression, "You never learn anything while you are talking," is a fairly valid statement. Listening to people express themselves is often a good way of getting to know them. Hopefully, a good supervisor will spend the greater portion of interviewing time being a good listener.

To understand a person's problem and to be offered intimate thoughts, rather than superficial feelings, requires positive, attentive listening. The listener must not jump to conclusions or make assumptions. It is not a good listener who thinks of what to say next, rather than paying full attention to the speaker. Rarely do individuals reveal everything at the beginning of an interview. Issues which come to the surface must be captured and analyzed; they may be expanded or resolved during the conference or at some later meeting.

Employees like to let off steam. Let them! This type of ventilation is a good device and can be used to facilitate revealing comments. Make guiding statements to effect a deeper exploration of sensitive areas of concern. Remember that personal feelings sometimes interfere with good listening and should be controlled.

Elements of Good Interviewing

An interview is different from a conversation in that there are, or should be, logical, straightforward reasons for conducting an interview, whereas a conversation can be held on the spur of the moment with no particular goal except sociability. The usual objectives of the interview include: (1) giving information, (2) receiving information, (3) motivating improved performance, and (4) counseling a person with a problem.

The most effective interview is often one that employs a nondirective interview approach. This does not mean that the interview has no approach; it means that the direction comes from the interview rather than being imposed upon the interview from an external source.

To illustrate this point, suppose that an interview is to be conducted to improve an employee's work performance. The supervisor might plan to deliver a lecture on the benefits of better work performance and might elicit promises from the employee to the effect that he or she will improve. This would be a direct interview approach. The result of such an interview probably would be that the employee would not improve, would resent the supervisor, and would lose much respect for the supervisor, who was satisfied with forced responses.

While it is easy to give an example of how the interview is externally directed, it is not easy to illustrate the correct method. The purpose is the same, but the method differs radically. It depends not only on the person involved, but also on the person's mood of the moment. If an employee comes into the office angry, it might be proper to let the rage run its course while listening sympathetically and then, casually, to introduce the subject of work performance. The supervisor will have to weigh sensitively the advantages and disadvantages of this approach because if the employee is extremely upset, it may not be possible to discuss anything reasonably. If, on the other hand, the employee comes angrily into the office and attempts to direct attention away from his or her shoddy performance—a sort of gambit where emotion is offered as a substitute for the problem—the supervisor should listen carefully to the tirade and then decline the game, thus avoiding complications. If the employee is friendly and cooperative, or confused and scared, totally different approaches are indicated.

The important point is that the design of the interview must emerge from the fabric of a human relationship. It cannot be painted on the surface. Any rigid interview plan the supervisor formulates is almost certainly doomed to failure because plans require some degree of predictability and people generally are not predictable. This should not be taken to mean that there are no principles to be followed or techniques to be learned.

Establishing Rapport

Rapport is one of those qualities that is easy to recognize when it exists, but difficult to describe. Through the years people have described the condition in many metaphorical ways. During World War II people would say they were "on the same wavelength," and young people a decade ago said that they were "grooving" with an individual. Both imply getting together and moving in the same direction, which is essential to the success of the interview.

While an officer, as an officer, might feel he or she has nothing in common with the interviewing supervisor, and vice-versa, if they discover that they are both fly-fishing enthusiasts an immediate bond is established. This then, is the first principle: establish rapport by finding some common ground that will enable both people to feel closer than two strangers. Incidentally, some imagination is called for; the supervisor who tries to establish rapport with everyone by asking about their children accomplishes exactly the opposite. No one likes to be treated as part of a formula. It is not that difficult to show a real interest in a person and, by observation, to discover a real common ground. Most people will reveal their interests by something in their dress, by some affectation, by their reading material, or in their casual conversation. The establishment of rapport allows the interview to proceed on the basis of two people who are interested in one another's well-being rather than on the basis of one member of the administration dealing with a subordinate.

There is a natural tendency for a supervisor to dominate a conversation, forgetting that the employee should be the central figure. Remember the objective of the interview. Arguing or finding fault will block effective conversation. A certain amount of rationalization will develop from time to time. Expect this and at the same time use caution about becoming personal in a derogatory fashion. Never attack the person; attack the problem.

Keep It Private

The conference should be private. Both persons should be comfortable so that there is a relaxed atmosphere. Imagine the impact on an employee if he or she were being observed from behind closed glass doors. When a supervisor is interviewing a subordinate, discretion must be used and all matters must be kept confidential. If the employee gives information that compromises the supervisor's duty, that information may be the object of real concern. A good supervisor must always consider the consequences of any statement that obligates the person to hold in confidence matters that involve possible violations of the law, major breaches of the department's regulations, and the like. Once the proper rapport has been developed, an employee who suggests that he or she will entrust a supervisor with certain knowledge which the supervisor must never reveal must be advised that if that information conflicts with the supervisor's duty, the supervisor cannot receive that information. However, the employee should be encouraged to divulge the information and assured that proper action will be taken in accordance with the responsibility that each has to the police service. Fair play and honesty are crucial. If it is not essential to department discipline, the supervisor should not divulge the information or discuss it with anyone.

Lead Employees to Solve Their Own Problems

The supervisor's contribution to the solution of the problem should be limited to those areas mentioned. He or she should be influential rather than persuasive. Sometimes the supervisor might be able to rephrase the problem or proposed solution and subtly modify it. Even hearing the problem in someone else's language might give the employee some new insight into it. Rephrasing the problem also forces the employee to do some reflecting, which usually leads to a somewhat different view.

At times there is no alternative to giving advice. If the employee refuses to face the problem, is hopelessly bewildered, or simply demands that the supervisor give an opinion, the supervisor has no choice but to offer some advice. In these cases the supervisor still has one recourse: the individual can be advised to see an expert. If the problem under attack is personal, and especially if it pertains to marriage, the supervisor is making a grave error in judgment in not sending the employee to a qualified expert.

If the problem pertains to the job, the supervisor should be able to help but may have to seek technical assistance. This procedure offers a double benefit: not only does the supervisor avoid the possibility of giving faulty technical advice based on inadequate information, but a team spirit is also fostered. The supervisor sets an example for the subordinate by relying on a qualified member of the team for help.

Solving problems in the manner described—establishing rapport, helping the employee analyze the problem, encouraging him or her to choose a solution, and, if necessary, soliciting help from an expert—is not only efficient but also gives the employee practical instruction in the *process* of problem solving. Should the employee someday become a supervisor, the example will be used as a model.

Avoid Emotional Involvement

The interviewer must avoid becoming argumentative or displaying authority. Admonitions and criticism are blockades in the road of communication. The supervisor who lets personal ego enter the picture or who becomes involved emotionally will not achieve the desired objective. There is no place in the proper interviewing atmosphere for the lightning flashes of ego or the storm clouds of emotion. The supervisor must check out the statements made whenever possible and must not take everything at face value. The amount of experience a person has often influences the interpretation of facts.

Special Interviews

The Induction Interview

The supervisor should arrange for a private, welcoming interview with each new recruit assigned to the unit. The meeting will help determine the recruit's attitude toward the supervisor when the person is in the most receptive frame of mind. This in turn will affect the individual's attitude toward the unit, toward the police service, and toward his or her role as a law enforcement officer.

The authors consider this interview to be an integral part of training, coming as it does at the very start of an individual's on-the-job experience, and it is discussed in more detail in the chapter on the supervisory training function.

The Progress Interview

The primary goal of this interview is a review of the professional progress of the employee. The strengths and weaknesses demonstrated by performance are discussed in hopes of eliminating the weaknesses. It is necessary to let all employees know how they are regarded with respect to performance. A real test of a supervisor's intestinal fortitude occurs when telling a subordinate that he or she is not performing satisfactorily.

The interview should begin with an explanation of its purpose, and there should be sufficient conversation to put the subordinate at ease. A good approach to the subject is to build the discussion around good job performance and to point out deficiencies in the work itself. An example of this approach is to say, "This is a poorly written report," not "You are a poor report writer."

If the employee agrees that the work is poor, let this person offer a suggestion for improving it. Encourage the employee to talk about the work; new areas may be developed. If he or she does not come up with the solutions, the supervisor must then make suggestions for improvements.

This is an example of a positive interview in which direct points are communicated. When closing this interview, the supervisor must encourage the employee by reference to his or her good points. The supervisor must help build confidence in the individual so that deficiencies can be overcome.

The Grievance or Complaint Interview

The reason for this interview is to discuss a complaint made by an employee or a complaint made about an employee. It is important to hear the whole story and, when appropriate, to work out a solution.

Make the person feel welcome. This type of interview is not usually planned, as the supervisor does not know when complaints will be made. Listening techniques are most important. If ventilation, or letting off steam, is all the employee needs, interruptions and arguments will destroy the entire interview. Remember to listen! Frequently, complaints are withdrawn after a successful interview.

In this type of interview, it is important to remember one thing: feelings can not be countered with fact. If an employee says, "I feel bad about having had to work overtime four days in a row when Jones didn't work any," this individual is telling the supervisor how he or she feels. And that is that. No facts are called for. Without doubt, the worst possible way to handle this type of grievance is to bring out the rule book and quote some inane regulation that says, "Employees will work overtime when required." The officer knows this or the overtime wouldn't have been worked in the first place.

When feelings are expressed, what is required of the supervisor—if anything—is feeling! If, in the example above, the supervisor answers, "I'm sorry. I know you didn't get to see your family much last week. And I really felt lousy saddling you with all the work," the interview to all intents and purposes is probably over. Some explanation might be offered, or the individual might be complimented for a job well done, but even that is extraneous. The employee wanted to communicate a feeling and the supervisor's response indicated that this feeling had been successfully communicated, so there was really nothing else to be said. Sympathy, appreciation, compassion, and understanding are the operative words here.

The supervisor must be able to differentiate clearly between this type of grievance and an outwardly similar type. If, in the example, the individual had said, "I worked overtime four days last week and Jones didn't work any. I don't see any reason for it and I don't want it to happen again," this person is asking for positive action and reassurances. Now facts are necessary and feelings are irrelevant. Differentiating between these two is impossible unless the supervisor is sensitive to the needs of the human being who is sitting on the other side of the desk.

When there is a basis for the complaint, the supervisor should take some type of follow-up action. A number of people may have to be contacted to determine the nature of the problem and the jurisdiction involved in seeking an adequate remedy.

When terminating this type of interview the supervisor should express appreciation to the employee for coming forth with the problem. Additional interviews may be required. Encourage the employee to bring in any data that will help to evaluate the matter. Tell the employee that he or she will be advised about the disposition of the case. The follow-through is vital to good communications between supervisor and subordinate.

The Personal Problem Interview

The interviewer's goal should be to help employees solve their own problems. Normally, the supervisor should not be involved unless the personal problem affects the employee's work. However, many problems are brought to the average supervisor's attention because of a close relationship with personnel.

Some of the techniques that encourage self-evaluation may be used here. The hazards of giving advice have been discussed. The supervisor can help the employee come up with a plan of action by guiding the conversation to those points that need consideration; the supervisor should not inject his or her own values into the discussion. It is certainly reasonable to motivate an employee by citing the elements of the problem so that the employee may seek the proper help. The supervisor should not hesitate to call attention to possible difficulties in the solutions presented by the employee.

Terminating the personal problem interview is not a simple procedure, since solutions are seldom clear-cut or immediate. Further discussions may be necessary. In compensation, the relationship that results from a successful interview will have many residual values.

The Disciplinary Interview

The goal of this type of interview is to secure improved performance with a minimum of dissatisfaction, resentment, and bitterness. Discipline should be thought of in terms of training and instruction rather than in terms of punishment and penalties. The subject of discipline is developed in depth in Chapter 10.

It is very important to future relations that disciplinary action be kept impersonal. The attitude the employee has toward the department is basic to both present and future changes in behavior. The punishment should fit each case; the key is fairness. The techniques involved in handling complaints against personnel are discussed elsewhere—the purpose of this interview is to make a disposition of the matter.

In planning the interview, the supervisor must be sure that all of the facts are in the record. Some additional homework may be necessary to insure a successful interview. Previous actions, rating reports, incident reports, and other factors come into play at this point.

Each supervisor should consider his or her own attitude toward the subordinate. A person should never interview when angry, even if it means delay, unless there is some emergency.

Privacy is an absolute requirement. It is very poor judgment to let others know that a subordinate is on the carpet. To set up a public display of the procedure is a grave error.

An opening remark should establish the purpose of the interview. Never start with accusations, but rather by advising the employee, "I'm here to talk to you about such and such. Do you want to tell me about it?"

If the supervisor has not participated in any previous investigation of the material at hand, the points to be emphasized should be considered. The subordinate in many cases will cover fully most of the points. Listen carefully. If a misrepresentation is apparent, let the employee know about it without encouraging argument. Review the facts and ask for any information that the employee feels was not included, such as any justification for the matter. Make a decision! If the supervisor must refer the matter to a higher authority for a final decision, emphasis should be placed on future improvement. The issues involving dispositions or recommendations are covered in Chapter 10.

It is good to close the interview on a positive note. If the supervisor feels the employee will improve, he or she should say so in simple language. Do not let the interview destroy the subordinate's self-confidence, but let the individual know that a responsibility to the organization exists. The matter should remain closed unless there is another violation; not to assure the employee of this fact is to miss a good bet for positive future relations.

The Termination Interview

The goal of this interview is to discover the employee's opinion of working conditions when the individual is leaving for reasons other than disciplinary problems. When a subordinate transfers, or is promoted and moved to another division, the same technique applies. The object is to determine how the person feels about working conditions in order to improve them.

Some planning, including the accumulation of background information, is helpful. The supervisor should create an atmosphere that will put the employee at ease and encourage the person to talk about the job. Some supervisors have been able to retain good individuals who were leaving by solving small grievances that had been bothering them. The personnel department should not discover problems that a good supervisor should have known about. Conclude the interview on a positive note. Record the items that were developed in this interview. Consult those concerned and, if anything can be done, start the ball rolling.

DISCUSSION QUESTIONS

1. What is the first rule for supervisors when talking with subordinates?
2. What are the elements of good interviewing?
3. How should the supervisor respond when an employee offers to entrust him or her with information that must not be revealed to others?
4. How should the supervisor handle requests for advice in personal matters?
5. How should the supervisor end a progress interview in which he or she has discussed an employee's shortcomings?
6. What is the purpose of the termination interview?

13 Complaints and Grievances

The proper handling of complaints and grievances, along with an active program for foreseeing and preventing them, is necessary in police supervision to prevent the undermining of employee morale by unresolved problems. The logical person to handle employees' grievances is the immediate supervisor. Grievances should not be regarded as a nuisance, but as an opportunity to correct an unsatisfactory situation. The supervisor should try to look upon a complaint as a suggestion; in many cases, the only real difference between the two is the manner in which they are communicated. A complaint or grievance can be aggravated if it is handled poorly, but a good supervisor can become established as a leader if complaints are dealt with effectively.

It is vital that the supervisor handle grievances, misunderstandings, and other employee relations matters promptly. To this end, the supervisor should have a thorough understanding of departmental policy and procedure relating to the usual situations that arise. If the supervisor lacks the necessary information, he or she should know, or be told where to find it.

Definitions

Grievance

A grievance is any matter causing discontent or dissatisfaction arising from the job that the employee thinks, believes, or feels is unfair, unjust, or inequitable. It does not matter if the grievance has been expressed or if it is valid. What is important is not whether the employee is right or wrong but the way that the employee *feels*.

Complaint

A complaint is the simplest sign of a grievance brought to the attention of management. It may be spoken or written.

Dissatisfaction

Dissatisfaction is anything that disturbs an employee, whether expressed or not. Dissatisfaction may be expressed in poor work performance, tardiness, and in other ways that will be discussed later in the chapter. General dissatisfaction might be caused by several specific grievances.

Personal Problems

A personal problem is an irritation or misunderstanding that, in general, does not arise out of the work situation as such. Personal problems may originate in the employee's home, with outside interests, or in interpersonal relations on the job. Clearly, there is some overlap between the definitions of a grievance and a personal problem. To clarify the distinction, if an employee had difficulty working with another individual because of ethnic background, the employee would have a personal problem; if the employee had difficulty because the individual would not carry an appropriate share of the work load, the employee would have a grievance. As a general rule, supervisors are concerned with personal problems only when those problems in some way affect job performance.

Prevalent Sources of Complaints and Grievances

Work Assignments

Work assignments may cause dissatisfaction and generate complaints. Some of the more common grievances are:

1. Level of work below the employee's capability and outside of his or her area of interest.

2. Poor or unfair distribution of rush and undesirable jobs. The continuous "crisis" or "panic" situation that requires extra work is symptomatic of poor planning and supervision and often results in grievances.

3. Inequitable distribution of overtime work. This includes "extra" work as well as regular police-duty overtime.

4. Frequent interruptions of assignments. Reassignment is especially resented when caused by poor planning.

5. Jurisdictional conflicts.

6. Basic environmental working conditions, such as poor lighting, wide temperature variation, inadequate ventilation, and poorly maintained or inadequate equipment.

7. Inadequate job instruction.

8. Work that is inconsistent with the grade classification.

9. "Siberia" type duty assignments.

10. Unfair methods for selecting men and women for preferred assignments.

While these do not represent all the possible grievances that an officer could have pertaining to work assignments, they are most of the more common ones. The supervisor should examine the methods utilized to assign work, should look at the conditions surrounding subordinates in the light of these common grievances, and should take the steps necessary to make sure that subordinates do not have to make similar complaints.

Attitudes, Actions, and Complaints

It is very important that supervisors control their own behavior, as their behavior influences subordinates and may either foster or help minimize grievances. Complaints and grievances often grow from supervisors' actions and attitudes; sometimes even the smallest act may cause a misunderstanding. The only means of preventing this is to have each individual in the organization be aware of the impact of his or her personality upon the people with whom he or she associates.

Complaints and the Professional Police Officer　　If the police officer's outlook is truly professional, the officer desires prestige and acknowledgment of his or her status. Supervisors should be considerate of this need in daily contacts. The use of plural forms—"we," "our," "us"—in communication is a small but important mark of the professional. Supervisors should share the credit for work in which an officer participates. Good supervisors share the recognition for good work and assume the responsibility for poor work and for poor unit effort.

The professional employee needs, among other requisites, some opportunity for self-expression and precise answers to questions. A professional usually does not like to take complaints outside the unit. Nor does he or she like to bother a supervisor with personal problems. As a consequence, the individual may keep small complaints bottled up inside until they accumulate to the extent that an explosion occurs.

The Supervisor's Attitude The following list gives some of the actions and attitudes which should be avoided in dealing with subordinates. Many have been mentioned elsewhere in the book, but they are listed here because they are the primary factors that determine whether a supervisor will get many complaints from subordinates or will get relatively few.

1. Failure to communicate
2. Oversupervision and unreasonable work standards
3. Failure to delegate responsibility and authority
4. Failure to relinquish detail work
5. Incorrect application of aggressiveness—needling employees
6. Restraint on employee's self-expression
7. Criticism of an employee in front of fellow workers
8. Failure to recognize meritorious work
9. Taking credit for the ideas of subordinates
10. Passing the buck
11. Disinterested attitude toward employees
12. Conferences or meetings called without adequate notice, planning, or outline of topics to be covered
13. Favoritism
14. Unprofessional attitudes such as bias, prejudice, and discrimination
15. Too much pressure; not enough pressure
16. Failure to set a proper example
17. Ignoring the errors of subordinates
18. Lack of support—failure to back up subordinate employees
19. Failure to hold subordinates responsible for their statements and actions
20. Arbitrary actions or decisions
21. Lack of counseling and interviewing skills in personnel matters

Attitudes and Actions among Employees Just as the supervisor's attitude will affect the conduct of subordinates, so will the attitude of each of the employees affect the conduct of the others. A person who is uncongenial or antagonistic is bound to make fellow workers miserable, producing an atmosphere conducive to grievances and complaints. Employees who make a fetish of their badge, becoming "badge-happy," can be extremely irritating to their fellow workers. Even a very aggressive attitude toward work assignments can cause serious problems.

The supervisor must continually watch for employees who are creating dissatisfaction among their fellow workers and take immediate steps to neutralize the situation. The effectiveness of an entire group can be reduced by one such person. Sometimes the offender can be reasoned out of destructive attitudes; at other times more drastic measures will have to be taken.

Compensation

Although compensation is not a prime motivating force for employees, it is still of considerable importance. If the pay scale is generally too low, dissatisfaction is a certainty and grievances will be almost a daily routine. Even worse is the situation that is usually found in small organizations where promotions and the corresponding salary increases are few and far between. Some individuals are naturally more ambitious than others and those men and women will either move up in the organization or cause dissatisfaction. If the organization makes no provisions for these people, it must be prepared to deal with their grievances.

The President's Commission on Law Enforcement and Administration of Justice published a report dealing with professionalism and career service which emphasized the need for upgrading the police officer. This report, *Task Force Report: The Police,* should be required reading for all supervisors.

Employees can easily compare the type, amount, and difficulty of the work that is required of them. If the pay differentials are unfair, there will obviously be problems.

Some of these problems cannot be solved by first-line supervisors, but the supervisor must be able to answer the questions and explain the pertinent policies to subordinates. Some of the problems can be solved by taking them up the chain of command; when serious dissatisfaction exists, it is absolutely necessary that the supervisor do this. Some of the concepts outlined in the *Task Force Report,* mentioned earlier, offer worthy goals for the future in the area of incentives.

Recognizing Dissatisfaction

It is always a good idea to anticipate problems and deal with them before they become really serious. A supervisor often has a chance to deal with an employee's problem before being presented with a formal complaint. All that is required is that the supervisor be observant. A certain degree of sensitivity is also helpful. A person with a grievance will usually reveal by his or her manner that something is amiss. Low morale is a good indicator of general dissatisfaction. The following list contains characteristics which should be regarded as warning lights, signalling that something is wrong. The conscientious and observant supervisor will heed these signals and deal with the situation before it becomes really serious.

1. Lack of cheerfulness or enthusiasm
2. A critical or antagonistic attitude toward the job, the department, the supervisor, or fellow employees (often displayed by disparaging remarks)
3. Loss of interest in work
4. Sullenness or surliness
5. Excessive tardiness and absenteeism
6. Low work output
7. Too many errors
8. Lack of cooperation
9. Neglect of responsibilities
10. Reduced aggressiveness and initiative
11. Nonobservance of rules and regulations
12. Heated arguments or quarrels with fellow officers
13. Evasiveness
14. Taking advantage of privileges
15. Requests for transfer
16. Neglect of appearance and equipment
17. Subversive behavior associated with disloyalty and conflict with management goals

Forestalling Complaints

In the last section, methods of recognizing dissatisfaction were discussed. The question now is, what should be done? It is impossible to observe the warning signals from behind the closed door of an office, so the first step is obviously to talk to the employees and be observant. It is helpful to go through a mental review of each officer each time the person is met. A set of questions can be memorized to facilitate this review. Is the individual's work satisfactory? Does the officer seem interested? What is his or her attitude? Does the individual seem to be cooperating with the other officers? Answering these questions on each encounter with an employee will force the supervisor to be observant.

If the answer to any one of these questions is negative, a more prolonged talk with the officer is indicated. If something seems wrong during this talk, the employee should be asked to the office or some other private place for a more serious discussion. The supervisor should then tactfully but determinedly find out what the problem is and ask how assistance can be provided to solve it. The supervisor should develop an interest in the officers. A sincere interest will prove very helpful when dealing with problems and complaints. Feigned interest is worse than none at all and may cost the supervisor considerably in terms of prestige.

The following are some additional suggestions for forestalling complaints:

1. Inspect working conditions periodically.

2. Help subordinates to improve themselves and get ahead.

3. Discover what status symbols are important to each individual and try to provide them.

4. Attack problems vigorously. Show interest in the employee's welfare.

A Plan of Action for Handling Complaints and Grievances

A plan of action for handling complaints and grievances is essential. The supervisor must not wait for problems to arise before deciding on a course of action. A good technique is to simulate various problems and then decide how each problem should be handled.

Initial Interview

A supervisor always must consider the basic objectives of the interview. The following suggestions and reminders will make the initial interview more useful:

1. Be receptive and available.

2. Handle the complaint as soon as possible. Set a time and place for the interview if duties prevent an immediate conference.

3. Hold the interview in complete privacy, and keep interruptions to a minimum.

4. If the employee is emotionally disturbed, do not respond to this mood. Try to be calm, speak slowly in a low voice, and be friendly.

5. Let the officer state his or her case. Do not interrupt except to ask questions that will help the officer tell the story.

6. Avoid introduction of your own bias.

7. Do not make a decision at the first conference if you do not have all the facts of the case or are not authorized to make a decision.

8. Tell the employee when he or she may expect to hear from you, and be sure that you do contact the employee.

9. Let the individual know that you welcome further discussion of the matter.

10. Do not overlook the possibility that the employee needs to blow off steam; it may dissolve the grievance.

In some cases it is beneficial if the supervisor uses the role-playing technique to provide a means for self-analysis through transfer of identity. For example, if an officer is disturbed because a supervisor refused to grant a request, and if the refusal was based upon valid grounds, the supervisor should ask the subordinate what action the subordinate would take if he or she were a supervisor and a subordinate made a similar request. In most cases, the subordinate will agree that the supervisor's actions were correct. The officer will also gain greater insight into the supervisor's role.

Gathering Facts

The supervisor must get all the facts from all the available sources. These sources include:

1. All employees concerned
2. Personnel records
3. Department policies, customs, and rules and regulations
4. Conditions and methods of work involved

Determination of Jurisdiction

The supervisor decides under whose jurisdiction the complaint falls and should not make any decision if the complaint involves one of the following:

1. Employees not under immediate supervision
2. The expenditure of more money than is authorized
3. A change in department policies, rules, or regulations

4. A change or modification of department-established standards of procedures

When a complaint is sent to a superior, the supervisor submits a statement of the facts of the case. Others are not committed to a course of action which cannot be followed.

Advice

The supervisor can seek advice of the personnel staff when it seems desirable and can consult an immediate superior for advice and support.

Decision Making

Decision making involves the following steps:

1. Assemble the facts in an orderly arrangement.

2. Study the facts and list all of the possible solutions.

3. Test each solution by questioning its feasibility. Does it conform to department policies and rules and regulations? What effect will it have on employee discipline, efficiency, and morale? Will it contribute to a smooth-running organization?

4. Select the solution which best meets the above tests. When in doubt, the solution can be checked with the supervisor's superior.

Taking Action

The supervisor should make the changes which are desirable and practical under existing conditions. He or she should confer with the officer, tell the person what has been done, and sell the officer on the fairness of the solution, if possible. It is important to establish a friendly relationship so that at the close of the interview the officer will leave with a desire to carry on as a working member of the team. If the officer is not satisfied with the solution, the supervisor should advise the person about other approaches to the problem which remain open.

Follow-up on the Action Taken

The supervisor's follow-up includes the following:

1. Live up to commitments scrupulously. Do not make commitments that cannot be fulfilled.

2. Talk to the employee after a few days to find out his or her current opinion of the situation.

3. Observe the person at work and go through the mental review described earlier.

4. The supervisor should use the experience for self-improvement. Could he or she have handled the situation better? Could the complaint have been anticipated?

It should be recognized that in any organization it is impossible to prevent all grievances. It is not only impossible but undesirable to promote a climate in which employees would be reluctant to voice their complaints for fear of some threat to their security. Rather than criticizing the employees for their gripes and complaints, supervisors should encourage employees to offer constructive suggestions.

The logical individual to handle complaints and grievances is the immediate supervisor. This person is in a position to detect the grievance in its early stage of development and knows—or should know—the workers. The way the supervisor approaches the grievance will either aggravate the situation or result in an amiable and equitable solution. In many cases, the supervisor who settles grievances is getting something from subordinates rather than getting something out of them. The supervisor is doing the dual job of representing subordinates to management and representing management to subordinates. Also, the supervisor is becoming established as a leader.

No two grievances are alike, so they must be handled according to the principles which have been discussed in this chapter. Frequently, it is not so much what is done but the manner or method of doing it that is important in resolving grievances; good judgment and the proper approach are essential. The supervisor always plays an important role in determining both the number and type of grievances in the unit. The effective supervisor will treat the grievances of subordinates in the same manner in which he or she would like to be treated. The mark of a good supervisor is not how well he or she is able to handle trouble, but how little trouble the supervisor has and how well he or she is able to forestall trouble by good supervisory practices.

DISCUSSION QUESTIONS

1. Discuss some of the psychological needs of the professional police officer that affect job satisfaction.
2. What are some of the ways work assignments may cause complaints or grievances?

COMPLAINTS AND GRIEVANCES

3. How can the attitude and actions of a supervisor influence employee discontent?
4. What are some of the indications of a dissatisfied employee?
5. How may a supervisor forestall complaints?
6. Discuss some of the ways a supervisor should handle a complaint interview. What use can the supervisor make of the role-playing technique?
7. Under what circumstances should a supervisor make no decision on a complaint?
8. What are the elements of the decision-making process in a complaint case?
9. How should the supervisor follow up on action taken?
10. Why is the immediate supervisor the logical person to handle complaints and grievances?

14 Decision Making and Planning

"Decision-making sustains bureaucracies, dominates legislatures, preoccupies chief executives, and characterizes judicial bodies. Decisions lead to policy, produce conflict, and foster cooperation."[1] Police officers, particularly supervisors, must make decisions and take action on matters that vitally affect other human beings. Frequently the situation demands that the decisions be made in a very short period of time. The decision made by the officer in the field will probably be "second-guessed" at many different levels over a considerable period of time—years if the second-guessing is being done by the appellate courts. The police officer's decision, therefore, must be right most of the time or the officer and the agency will be subjected to a great deal of adverse criticism.

The Importance of Supervisory Decision Making

The success of contacts between police officers and citizens depends largely on the manner in which the officer takes action. This action is the direct result of decisions made by the officer which, in turn, must be within the framework of the law and the agency's policies. Inasmuch as a supervisor must accept full responsibility for the decisions of subordinates, it is the supervisor's duty to provide these subordinates with adequate knowledge of the agency's policies and procedures. In addition, the supervisor must be prepared to make many important decisions personally and to accept complete responsibility for their consequences.

The importance of supervisory decision making is reflected in the literature of psychology, business administration, sociology, economics, public administration, and many other disciplines.

DECISION MAKING AND PLANNING   189

An example of the interest in the subject is Paul Wasserman's and Fred Silander's *Decision-Making: An Annotated Bibliography,* which covers the period from 1945 through 1957 and contains 438 entries. A second edition, for the period from 1958 through 1963, contains more than 600 entries, despite the fact that greater selectivity was used in the second volume.

"Whatever a manager does, he does through making decisions. Those decisions may be made as a matter of routine. Indeed, he may not even realize that he is making them. Or they may affect the future existence of the enterprise and require years of systematic analysis. But management is always a decision-making process."[2] "There has been a general recognition of the importance of decision-making in supervision but a good deal of the emphasis has been placed on the solutions, with little stress on the proper definition of the problem. Decision-making requires first a definition of the problem; only then can a solution be sought."[3]

Elements of Decision Making

Important decisions, those that are most strategic, depend upon properly identifying existing conditions and then changing or modifying them. There is no standard formula that assures the proper decision in all cases, but it is hoped that the procedures outlined in this chapter will assist the supervisor in applying sound judgment when making decisions. A distinction should be made between the strategic decision and the routine, unimportant decision that is chiefly an exercise in problem solving. If both the conditions of a problem and the requirements that need to be satisfied by the solution are known and simple, the decision is only a matter of choosing between a few obvious alternatives. Usually the decision that will be made accomplishes the desired end with the least possible amount of effort and disturbance of routine affairs. For example, in deciding which of two radio dispatchers should get coffee for the radio room, a simple question is, "What is the prevailing etiquette or custom?" A more complicated question is, "Should there be a coffee break during the tour of duty and, if so, does the break result in a gain or loss of work accomplished? And, if the loss outweighs the gain, is it worthwhile to upset an established custom in the department for the sake of a few minutes?"

However, determining the right questions and answers is only the first step toward solving the problem. It is even more important, and often more difficult, to implement whatever course of action has been decided upon. A police agency cannot be concerned with knowledge for its own sake; it must be concerned with knowledge that contributes to the performance of the task. For this reason, there is nothing quite as useless as an effective solution that is put into a filing cabinet or ignored by the people who are supposed to act on it. One of the most critical requirements in the decision-making process is that the decisions reached by different individuals in different units of the department be compatible with each other and with the policies of the agency.

Public Nature of Police Decisions

Because of the regulatory nature of police work, which imposes restrictions on the individual's freedom of action, decisions made within a police agency are always subject to close scrutiny by the public and the press. Bad decisions almost invariably result in unfavorable publicity for the agency as well as for the individual involved. All decisions made by police supervisors either become or affect department policy or procedure. According to Allen Bristow and E. C. Gabard,[4] if policies and procedures are clear-cut and can be easily understood by subordinates, police action becomes more effective. If policies are not clearly defined, preferably in writing, decision making becomes exceedingly difficult at all levels of operations.

Indecision

Indecision causes anxiety and frustration for both the decision maker and subordinates. Subordinates have no confidence in a supervisor who is not decisive—a disastrous situation in a police agency when orders must be followed without question or hesitation. Bristow and Gabard list the following results of indecision:

> The usual pattern of the inability to make decisions is as follows: the administrator, due to fatigue or nervousness, experiences indecision; he finds that concentration, beyond a brief time-span, is difficult; as a result of this realization, he tries to compensate for his inabilities by putting forth more effort; this merely creates tension and irritability; he must then compensate for these new factors; he experiences impulsiveness, and this, he finds, increases his burdens; realizing that he is becoming incapacitated, he makes quick decisions rather than attempting to produce accurate, well-thought-out decisions; this creates a sense of guilt, which in turn adds indecisiveness to future decisions, thus tempting the subject to avoid decisions, and convincing him of his own unworthiness; insecurity results. Usually, the administrator does not really comprehend the process, so he places the blame for his inabilities on his own shoulders rather than on the snowballing effect of the above process.[5]

Peter Drucker divides the decision-making process into five distinct phases: defining the problem, analyzing the problem, developing alternate solutions, deciding upon the best solution, and converting the decision into

DECISION MAKING AND PLANNING

effective action.[6] Other methods have been developed by other authors using somewhat different terminology (see An Outline for Decision Making, Figure 8), but the main elements of decision making are, nevertheless, fairly uniform.

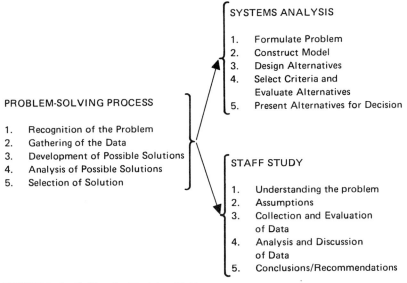

FIGURE 8. An Outline for Decision Making

Preparing to Make a Decision

The sources of the problem must be determined before a decision can be made. Many of the supervisor's problems are related to the job: the review of records and statistics, personal relationships, and referral of problems from subordinates or superiors. Other problems may originate outside the agency with individual citizens or organized groups that represent public or special interests. Before supervisors can make decisions, they must determine that the problem is their own responsibility. Some questions can and should be referred to other persons.

Find the Correct Decision Maker

The supervisor should not spend time on decisions that should be made at a lower level. The person must learn to trust subordinates with the responsibility for such decisions and must fight the temptation to do all the work personally. On the other hand, a supervisor should also avoid the urge to overdelegate authority.

Define and Clarify the Problem

The second step in decision making is to define and clarify the problem, which generally requires a consideration of the law and the agency's policy, as well as the use of good judgment.

The supervisor must look for the *key* or *critical* factor of the problem—the single element that, when changed, allows other factors to be changed or corrected. Too frequently, the decision maker begins to hunt frantically for answers before knowing which aspect of the problem actually requires a solution. When the supervisor is considering possible solutions, he or she must eliminate personal desires and concentrate on what is most beneficial to the department. This is more easily said than done.

Getting the Facts

The next major step in decision making is to get all the facts, particularly those regarding the critical factor of the problem. In the example in which the supervisor had to decide which of two radio dispatchers should get the coffee for the radio room, the critical factor was the question of whether or not there should be a coffee break during the regular duty tour. To answer that question, it is necessary to pose other questions. Is there a state law prescribing rest breaks during working hours? What is the established departmental policy if there is no law? Inadequate information usually results in poor decisions unless, on rare occasions, the supervisor is very lucky.

Time for Fact Finding

One of the big problems that confronts the average police supervisor is the amount of time allocated to fact finding before making a decision. Probably the best rule is to stop looking for facts when you're reasonably satisfied that you fully understand the situation. The supervisor will not be right all the time, but nobody is. It is important to realize that if the supervisor waits too long the situation may change before a decision is made and action is taken.

Sources of Facts

Facts may be gathered from many different sources. Individuals can often provide information pertinent to the problem. This information may be directly related to the current problem, or it may be related to the experience of a person with a similar problem. Both types of information are valuable in making a decision.

Public and private records and statistics are another valuable source of information. Records in the police agency's files are worth exploring, particularly if the same or a similar problem has been resolved in the past. Sources outside the agency must also be considered, such as the records of other agencies, journal articles, newspaper files, citizens and community groups, public libraries, questionnaires, and many others.

Objectivity of Information

The supervisor must make every effort to be objective when evaluating information so that personal opinions do not influence the interpretation of data.

Factors Limiting the Selection of Alternatives

The decision maker must not immediately accept the most obvious solution to the problem. There may be many alternative solutions that are not quite so apparent, and the process of elimination cannot begin until they are discovered and evaluated.

The Law

One of the most important factors that influences decision making in the police agency is the law, because it limits solutions to those that are within it. Therefore, the law should be consulted before any decision is made.

The Police Agency's Policy

Each solution must be examined to see if it is compatible with the policies of the agency or if it will establish a new policy. If a supervisor has the authority to establish policy, inconsistent, "this-time-only" decisions must be avoided. There can be serious repercussions when bad precedents are established.

The Individuals Involved

When it is possible, the individuals who will be directly affected by a decision should be consulted. They may be able to suggest a solution that the supervisor has overlooked. Also, it is much easier to make a change when the subordinates have participated in the decision-making process.

Special Interests and Politics

The possible effects of each solution on special interest and political groups should be calculated. Local politics vary considerably. In some cases, the agency must depend on local politicians for its operating budget and cannot afford to offend this group. While a supervisor's job may not be jeopardized by offending the city council, a proposed pay raise for the department may be endangered by precipitous action. If a special interest or minority group is offended, it can be extremely vocal in its criticism of the agency. Whether or not such criticism is warranted, the result is likely to be the same.

Morale and Discipline

The effect a decision can have on morale and discipline is another important factor to be considered. A decision that is not well accepted creates dissension, unrest, and low morale. An unpopular decision may be ignored, conveniently forgotten by subordinates, or filed in the "circular file."

Public Impact—The Press

The supervisor must attempt to predict the effects of each considered solution on the public, especially the way in which each will be interpreted by the press. It is easy to arouse public resentment, but difficult to placate it.

Finances

All police agencies must work within their budgets, so possible solutions will have to be discarded if they are too costly.

Making the Final Decision

After gathering all of the available facts—the supervisor will rarely be able to get all the facts—one solution may be an obvious choice. It is not necessary to have all of the available facts in order to make a reasonably sound decision, but it is necessary to know what information is missing in order to calculate the risk the decision involves and the degree of precision and rigidity that the proposed course of action can afford.[7] Various methods can be used to weigh the factors involved in making a decision, but all of them are subject to error because of the human element. However, there are not yet any machines or computers that are capable of thinking. The final problem of judgment is in the hands of the supervisor.

Avoid Indecision

It is possible that after weighing the considerations carefully each of the solutions will appear to be equally desirable. When this happens the supervisor may be tempted to make no decision at all, especially if the problem is not a crucial one. However, many decisions must be made promptly or important operations will be placed in jeopardy. In a few cases, a subordinate's job may be suspended until a decision is made, because if it is not suspended, it will be done in a haphazard manner, which causes confusion and bad morale.

While the public and superiors may criticize a supervisor for a poor decision, they will certainly criticize this individual for making no decision. Occasionally, however, the proper solution is to make no decision. When the supervisor encounters this situation, he or she must explain to subordinates that no decision is actually the most desirable solution. They must understand that the supervisor has decided not to act and realize that the problem is not indecision on the part of the supervisor.

Fast Decisions

Fast decisions are required most frequently at the lower levels of the police agency, usually in a field situation. Making a fast decision should not present any real problems, and the supervisor should not have to "play hunches" if properly prepared in advance.

Contingency Planning

The supervisor should anticipate various situations and then plan the action to be taken. A knowledge of decisions that were made in similar situations can serve as a guide but can never replace good planning. The department's training program should include the most common problems a police officer faces and the correct action to be taken for each one.

This sort of drilled reflex can reduce or eliminate the need for an on-the-spot decision when the situation actually arises. But the anticipated situation used for training must be one that admits of little or no variation. Otherwise an automatic response might be inappropriate, dangerous, or unjust.

Factors Influencing Decisions

Decisions can be profoundly influenced by factors unrelated to the original problem, such as the mental and physical condition of the decision maker.

Physical Condition

Whenever possible, decisions should be made only when the supervisor is in good health and is well rested. It is often impossible to delay an important decision, but supervisors should realize that many of those problems that seem to demand an immediate solution can be delayed until a more favorable time.

Mood

Bad moods are likely to cause pessimistic evaluations of problems, while good moods can cause unwarranted optimism. In either case, the decision maker builds a distorted picture of the problem that will probably result in a bad decision. Before attempting to solve a problem the supervisor should be in a proper frame of mind. If not, problem solving should be postponed if possible. The average quality of decisions will be higher as a result.

Demeanor and Attitude

The supervisor should consider the effects of demeanor and attitude on both superiors and subordinates when making a decision. If the supervisor does not appear confident, others will lack confidence in that person as well. The individual must be decisive and avoid vacillating between alternatives. A supervisor should be calm to avoid the appearance of having made a snap judgment, but, at the same time, should not appear lackadaisical. The impression should be conveyed that the supervisor has not been pressured into a hasty decision and that he or she is concerned about the outcome of the decision.

Implementation of the Decision

The people who will be directly affected by a decision should not hear about it "by way of the grapevine" because the information probably will be distorted by the time it reaches them. Subordinates in particular must have a clear understanding of what is expected of them. If the public is affected by a supervisor's decision, it should be informed before the decision actually takes effect. An example might be the situation in which jaywalking has resulted in a sharp increase in automobile-pedestrian accidents. A decision is made by a traffic supervisor to begin a rigid enforcement of ordinances against jaywalking. If a program of this type is sprung on an unsuspecting public it can only cause a quick deterioration of police-community relations and a great deal of resentment from those who receive citations. However, if the enforcement is preceded by a campaign to inform the public about the accident problem and the need for such a program, the public will be much more likely to accept the program when it is actually implemented.

Explain the Decision to Those Concerned

Because of the quasi-military nature of the police agency, there is a tendency to issue orders in a military style, without explanation. This practice can easily lead to a complete lack of cooperation on the part of subordinates, because they are not able to understand the reasons for the order. It is good policy for a supervisor to explain an order whenever it is practical. The need to know *why* applies equally to both police officers and the public. A failure to give any explanation usually results in suspicion and misunderstanding.

Timing; Feedback

Proper timing is another consideration in implementing a decision. Decisions should be announced at a time when all participants will be more apt to accept rather than reject them.

The supervisor must be prepared to justify and, if necessary, defend a decision until convinced that he or she has made an error. When an error is discovered, the supervisor should reevaluate the decision and be prepared to make a change if necessary. The supervisor should never be afraid to admit a mistake—everyone makes some. In police work, particularly, the failure to admit an error and to make the necessary correction can be disastrous.

Feedback is another important element in the decision-making process. It is often essential to build into a decision some system for information monitoring and reporting that will provide a means to evaluate its effectiveness.

People make decisions and people make mistakes. Even the most effective decisions will eventually become obsolete.[8] The supervisor should always try to build into his or her decisions organized feedback—figures, studies, reports—to monitor and return information on them. In spite of feedback, many decisions will not become effective, and thus, a willingness to correct errors is of extreme importance in decision making.

The good supervisor learns to accept full responsibility for decisions and their consequences. When things "go sour"—and they will occasionally—the supervisor should never "pass the buck."

Planning and Decision Making

Decision making and planning are closely related—a decision may be the beginning of a plan. Conversely, a plan may be the framework for decisions that follow. Thus, a supervisor should review past problems and anticipate future ones before a decision is made. The successful management and control of any large enterprise requires carefully prepared plans that can accurately forecast operations.[9]

Steps in Planning

The supervisor needs a plan to develop a method, procedure, or arrangement of parts that will facilitate the achievement of a certain objective. O. W. Wilson has developed seven fundamental steps involved in planning.[10]

1. Discovering the need for a plan
2. Formulation of a statement of the objective of the plan

3. Gathering and analysis of all relevant data

4. Development of the details of the plan

5. Obtaining concurrences from organizational units, both within and outside the department, whose operations may be affected by the proposed plan

6. Putting the plan into action

7. Analyzing the plan

Note the similarities between these steps and those involved in decision making.

Some plans are statements of goals—a request for better working conditions, higher salaries, or higher educational qualifications for officers. Other plans concern the routine procedures of a job—fingerprinting, report writing, and maintenance of up-to-date files. There are plans for special events—visiting dignitaries, parades, mob and riot control. Generally, for procedures that are not routine, plans cover one event only. However, they may well be used as references for future special events.[11] The supervisor who is responsible for planning should review all of the procedures that affect the unit to ensure that they are up-to-date and effective. The individual should suggest improvements and ask other supervisors for their cooperation in the formulation of new procedures when they are needed. Reports of operating units and data on crime and accident rates should be used to predict future needs.

The efficient police supervisor anticipates those situations he or she feels will frequently arise and makes advance plans to handle them. However, no amount of planning can predict all of the contingencies that will arise in a police agency's daily operations. For example, although a patrol supervisor can examine past reports and set up prearranged plans for dispatching units to business establishments that have been frequently robbed, it would be impractical to arrange such plans for *every* business establishment which *might* be robbed in a given district.

The good supervisor has developed to the utmost a capacity for making good decisions. He or she will plan ahead as much as is reasonably possible but will be prepared to make decisions as the need arises. Even those decisions that he or she makes rapidly in the field will be reasonably sound and well thought out, despite the fact that they are made quickly and under emergency conditions.

Other decisions, which do not have to be made immediately, will be carefully thought out and made only after proper evaluation of all of the alternatives. Good decisions are essential to proper planning, and effective planning eliminates the need for many decisions.

DECISION MAKING AND PLANNING

Management Supervisory Staff Studies

Competent and progressive law enforcement administrators are constantly seeking ways to better their department's responsiveness to community wants and needs, to clarify and refine goals and objectives, to maximize the utilization of agency resources, and to move capable personnel into positions of responsibility which are appropriate to the individual's abilities and skills. The police executive who seeks this maximum utilization of supervisory personnel is today greatly expanding the function and role of these persons within the organization.

Realizing that those closest to community needs and problems are often the best qualified to assist in the development of appropriate and responsive police services, the police executive will benefit the organization and the community by providing supervisory personnel with an opportunity to participate in the development of staff studies.

The supervisor should be aware that while such studies traditionally have been prepared by administrative "staff," regardless of capability, the modern team management approach to decision making dictates that the most capable individual for the job be given the assignment. Therefore, an individual in a supervisory position may be called upon at any time to research and develop a complex new program or to solve a difficult organizational problem.

The bottom line is that police executives need data and information upon which to base appropriate decisions or policy. There is nothing mysterious or complicated about such a staff study approach. In fact, it is a most efficient procedure, designed to use a disciplined, rational method of solving problems.

Guidelines for Preparation of Management/Supervisory Staff Studies

Police chiefs are routinely called upon to solve a wide range of problems demanding more time than is available to a single individual. One of the functions of the staff is to assist the chief in selecting and executing effective solutions to a myriad of problems. In order to obtain an accurate picture of a problem, the chief assigns it to one of the staff officers. Such an assignment is called a staff study. It requires that the staff officer (the "project" officer) research the problem area in detail in order to identify the key problem and recommend effective action based upon all the relevant facts and an accurate exploration of alternative solutions. In most instances, the completed study is submitted by the project officer to his or her superior in the form of a written report. Separate staff studies are sometimes conducted on component aspects of a larger problem. A staff study may sometimes be initiated and conducted by a staff officer who is particularly interested in a problem.

In some instances, the staff study assignment will come to a police officer in the form of a clearly stated problem. More commonly, it will come in the form of a generalized description of a problem or a set of related problems. The first step, then, is to identify and state the key problem. In order to insure that the study will be appropriate in direction and scope, a statement of the problem is submitted to the chief for executive approval. Following approval, the project officer's job is to seek solutions based upon relevant facts.

As the problem is studied, it will become clear that there is rarely ever one "obvious best" solution. Thus, all the reasonable alternatives must be considered and selection must be made of the one that, based upon judgment of all the available facts, seems to be superior. Since decisions always involve personal judgment, the project officer must be prepared to defend all decisions with documented facts and, where necessary, valid assumptions.

Usually, a project officer will be required to solicit concurring opinions from other staff members. Individuals who do not agree with the conclusions or with any aspect of the study are thus provided an opportunity to specify their nonconcurrence. When this happens, a section must be attached to the report in which the nonconcurrence is rebutted. But if the criticism cannot be rebutted, the entire study may have to be revised. The prudent procedure, then, is to anticipate nonconcurrences. This is done by determining possible sources of disagreement *during the conduct* of the study and by providing a firm factual basis for meeting those objections.

Part I: Identification and Statement of the Problem

A crucial step in the conduct of most staff studies is the identification of the specific problem implied by the general assignment from the chief. Most frequently, a rough description of the problem area will be given from which the project officer must determine exactly what it is the superior wants to know. Often, the superior may not be certain what the key problem is and may want the project officer, as a first step in the staff study, to determine the key problem. Unless the project officer is an expert in the area, some investigation will be necessary before that determination can be made. Once it is made, a statement of the problem should be drafted and submitted for approval before proceeding with the study. In order to be approved, the statement must name some specific action that describes the general purpose of the study—for example, to determine, to assist. Furthermore, the scope of the problem must be precisely indicated so that it is clear to both the project officer and to the chief exactly what is to be included in the study. This approval step will insure that the study will be neither too broad nor too narrow. Any subsequent changes in scope should be approved by the chief.

Part II: Research and Data Collection

Having identified an appropriate problem and received the superior's approval of the statement of the problem, the project officer begins the search for its solution.

The project officer should first engage in some "brainstorming," listing as many potential solutions to the problem as occur, without regard to their practicality. Having composed this list, the project officer should then begin intensive research to collect all the relevant facts connected with the problem.

The primary source of information for the study will be official documents. These include technical reports, operating manuals, previous staff studies, government reports, etc. This information may be gained from technical libraries, available bibliographies and abstracts, and other sources. If time permits, and if it seems appropriate, the project officer may supplement official document data by collecting original data from persons intimately connected with the problem. Included in this category are: (1) Experienced local colleagues, (2) Subject-matter experts, and (3) Operational personnel who have firsthand knowledge of the problem.

The methods for collection of original data include: (1) Interviews, by means of either telephone or personal visits; (2) Letter requests for specific information; and (3) Questionnaires directed to operational personnel.

Part III: Interpretation of Data

As data collection progresses, the project officer begins to pare down the list of potential solutions. He or she first rejects all solutions that, based upon the data, seem *unsuitable* to solve the problem. A solution is not suitable if it is not capable of solving the problem or accomplishing the mission.

Before completing the research, the officer informally seeks out those persons with apparently suitable solutions who are sources of potential disagreement with the officer's solution. These nonconcurrences are used to help guide further research.

Another guide to further research is provided by considering whether each *suitable* solution is also *feasible* and *acceptable*. Feasible solutions are those that can be implemented with available resources. Acceptable solutions are those that are worth the cost or risk involved in their implementation. This kind of decision suggests that further facts are necessary for complete evaluation of alternative solutions. Sometimes, relevant information is unobtainable. If information crucial to evaluating alternative solutions cannot be obtained, then it must be replaced by a *valid assumption*. Such assumptions should be considered tentative until data collection is completed.

The application of the criteria of suitability, feasibility, and acceptability thus helps direct further research in three ways:

1. By screening out unsatisfactory solutions,

2. By identifying solutions that should be checked for nonconcurrences, and

3. By calling attention to facts needed for evaluating alternative solutions.

Further research is thus directed toward facts needed to evaluate nonconcurrences and replace tentative assumptions with data.

Part IV: Evaluation of Alternative Solutions

After extensive research, the project officer must select the superior solution to the problem. He or she begins by screening out obviously unfeasible or unacceptable solutions. The remaining alternatives must then be weighed against each other singly and, perhaps, in various combinations. The superior solution is the one that best fulfills the criteria of suitability, feasibility, and acceptability. The solution chosen must measure up to all three.

If possible, the superior solution should rest solely upon documented facts. If a crucial fact is not available, it must be replaced by a *valid assumption*. Remember that valid assumptions describe conditions that must be fulfilled before the conclusions can be accepted without reservation. Thus, assumptions are to be avoided if possible. However, if a valid assumption is crucial to the project officer's decision, it must be stated.

In order to organize his or her reasoning, it is useful for the project officer to put ideas down on paper. This permits the officer, as well as outside critics, to evaluate the logic behind the report and point out any weaknesses that may exist before submission of a final report.

Part V: Preparation of the Staff Study Report

Staff study reports, while varying in format requirements from one organization to another, generally follow the format described below.

The body of the report contains—in three single-spaced typewritten pages or less—a summary of the information necessary for rapid evaluation. It contains these six basic elements:

1. Statement of the problem

2. Valid assumptions (if any)

3. Facts bearing on the problem

4. Discussion

5. Conclusions

6. Recommendations (if any)

The report may also include these elements:

1. List of annexes
2. Concurrences
3. Nonconcurrences
4. Considerations of nonconcurrences
5. Action by the approving authority
6. Annexes, containing, for example: detailed data, lengthy discussions, executive documents, and bibliographies.

DISCUSSION QUESTIONS

1. What is the relationship between decision making and policy?
2. What is the real importance of supervisory decision making?
3. What is the difference between decision making and problem solving?
4. How is the decision-making process related to problems of public relations?
5. What are the effects of indecision on the supervisor and the subordinate?
6. Describe each of the steps in decision making.
7. What are the origins of problems that require decision?
8. What factors enter into the evaluation of alternatives in decision making?
9. How do you select the best alternative?
10. What are some ways of gaining acceptance of a decision?
11. Explain the relationship between decision making and planning.
12. What are the different types of planning, and how is each used?
13. Why is a supervisor's participation in the preparation of management/supervisory staff studies important?
14. What are the basic elements of a management/staff study, and how is each developed and presented?

NOTES

1. James N. Rosenau, "The Premises and Promises of Decision-Making Analysis," *Contemporary Political Analysis* (New York: The Free Press, 1967), p. 195.
2. Peter F. Drucker, *The Practice of Management* (New York: Harper & Row Publishers, 1954), p. 351.

3. Ibid.
4. Allen P. Bristow and E. C. Gabard, *Decision-Making in Police Administration* (Springfield, Ill.: Charles C. Thomas, 1961).
5. Ibid., pp. 7-8.
6. Drucker, op. cit., p. 353.
7. Drucker, op. cit., p. 359.
8. Peter F. Drucker, "The Effective Decision," *Public Administration News Management Forum,* Vol. XVII, No. 2 (May, 1967), p. 2.
9. Donald C. Stone, *The Management of Public Works* (Chicago: Public Administration Service, 1939), p. 63.
10. O. W. Wilson, *Police Administration,* 2nd ed. (New York: McGraw-Hill, 1963), pp. 89-108.
11. *Municipal Police Administration,* 5th ed. (Chicago: International City Managers Association, 1961), p. 84.

15 Work Planning

The implementation of the police organization's procedure is a basic obligation of the supervisor. Since the effective use of both time and personnel is essential, the supervisor must be able to evaluate work performance, eliminate unnecessary jobs, and improve overall efficiency. The supervisor is, therefore, responsible for making work assignments that utilize personnel, equipment, and money in the most efficient way possible. In order to develop skills in work planning, the supervisor must evaluate his or her own work schedule to see if time is budgeted correctly. Only then can a supervisor begin to evaluate and adjust subordinates' schedules.

The Use of Time and Personnel

The reasons for improper use of time and personnel usually can be discovered by an analysis of the following areas:

1. Organization problems
2. Failure to delegate authority
3. Poorly selected or inadequately trained employees
4. Inadequate planning
5. Lack of appropriate allocation of time for specific tasks
6. Improper distribution of the work load
7. Improper deployment of available personnel

Utilization of Time

Some of the most common ways in which time is wasted—together with suggestions for improvement—are:

1. *Paper shuffling*—Make decisions promptly. Do not pass the buck if it is your job to make a decision.

2. *Slow reading*—Use different methods of reading for different types of materials (scan; read twice). Take a speed-reading course.

3. *Individual discussion*—Organize, plan, and summarize discussions.

4. *Conferences and meetings*—Set up an agenda in advance, including a tentative time schedule.

5. *Inadequate instructions to subordinates*—Plan and assign work properly. Anticipate questions and problem areas. Research problems when appropriate.

6. *Inadequate use of subordinates*—Delegate work with commensurate authority.

7. *Long telephone calls*—Be brief. Organize thoughts beforehand.

8. *Interruptions to determine work progress*—Set up checkpoints and expect progress reports. Condition subordinates to supply a status report. The need to brief higher-level supervisors is sometimes ignored or regarded as an inconvenience.

Utilization of Police Personnel

The following list contains some basic factors that affect the utilization of the work force. They should be evaluated by the supervisor.

1. *The individual police officer*—Training, experience, intelligence, personal qualifications, interests, knowledge, skills, attitudes, and physical condition. The amount of supervision required. The length of service with the department.

2. *The jobs to be performed*—The variety of jobs available.

3. *The working conditions*—Number of hours of work per day and per week. The physical facilities available.

4. *The immediate supervisor*—Effect of the supervisor on the police officer, on work assignments, and on working conditions.

The Use of Job Planning Procedures

Job planning is the division of work into jobs that utilize and develop most effectively the police officers' knowledge and skills. Tradition and chance are not effective; analysis and planning are required.

Why Job Planning Is Essential

> 1. *Characteristics of police work*—Diversity of a police officer's duties and responsibilities. Variety of skills required in different phases of police work.
>
> 2. *Characteristics of individuals*—Number of traits and abilities possessed by each individual. Traits of individuals are exhibited in different degrees and at different levels, and are independent of other traits.

Effects of Job Planning

> 1. *How job planning affects the police officer*—Makes an officer's job more interesting. Clarifies opportunities for transfer to desirable positions or for promotion.
>
> 2. *When job planning affects the police officer*—During recruitment. At the time the individual enters the police service. During employment. At the time of performance appraisals and promotions.
>
> 3. *How job planning affects the police agency*—Increases work output. Reduces cost. Improves employee relations.
>
> 4. *How job planning affects the supervisor*—Improves the individual's reputation with superiors, associates, and subordinates.

Steps in Job Planning

> 1. *Obtain information about individual police officers and their skills*—What can they do? What do they want to do?
>
> 2. *Obtain information about the jobs to be done*—Tasks performed by each person in the unit, and the time spent on each one. List of responsibilities and operations assigned to the unit. Work distribution chart, including tasks and operations.
>
> 3. *Attempt to reconcile individuals and jobs.*
>
> 4. *Appraise and re-evaluate original work assignments.*

Analysis and Study of Work to Be Done

1. *Are all necessary operations and tasks included in the analysis?*

Is the time spent on each task and operation appropriate? What operations and tasks take the most time? Is effort being misdirected? Is too much time being spent on nonessential details? Does each task contribute to the objective of the unit? Is each operation essential to the objective of the unit? Does any task unnecessarily duplicate something that is being done elsewhere in the unit or department?

2. *Are officers' skills being used properly?*

Are some individuals being assigned to tasks that do not utilize their skills? Do some officers have duties that are beyond their skills or for which they have not been trained?

3. *Are duties being spread too thinly?*

Are many officers doing a small or unimportant task that a clerk or other civilian employee could do equally well? Are some tasks given to so many persons that no one is responsible for them? Is the work distributed equitably? Does each individual carry a fair share?

4. *How will emergencies (illnesses, vacations, etc.) be handled?*

Can flexibility be developed by job rotation? Who knows how to do each job? When was it done last? Are department procedures kept up-to-date? Are they in written form?

Overall Job Planning

Some basic steps in overall job-planning procedures should be considered:

Study existing jobs to get the general pattern for each.

List the tasks and operations. Group operations by types into major divisions. Identify repetitive operations or tasks. Identify differences in type and level of skill.

Plan tentative job arrangements.

Check with those who know the work. Evaluate and choose the best job patterns. Test the chosen pattern and develop a new one if it proves ineffective. Observe and improve on the new job pattern.

Planning and Assignment of Work

What Is Meant by Assigning Work?

A work assignment should answer the following questions. What is to be done? Why it is to be done? Who is to do it? How is it to be done? Where is it to be done? When is it to be done? What is the estimated amount of time required to do it? It is necessary to assign work effectively in order to obtain the desired results from subordinates. Desired results include the quantity, quality, time required, and cost of the job to be performed.

Essential Elements of Proper Work Assignment

The essential elements of proper work assignment are:

1. Mutual understanding between supervisor and subordinate of the concepts of work assignment and delegation of responsibility

2. Obtaining preliminary information needed for planning the work assignment

3. Planning the work assignment

4. Making the assignment

5. Following through on the assignment

Concepts of Work Assignment and Delegation of Responsibility

The concepts of work assignment and delegation of responsibility are:

1. The delegation of responsibility does not relieve the supervisor of accountability for the job.

2. A supervisor is always responsible for the performance of subordinates.

3. A subordinate must be willing to accept responsibility and must be aware that the superior is responsible for the job he or she performs.

4. Assignment of responsibility should be accompanied by commensurate authority—the authority required to fulfill the responsibility.

Why Is It Necessary to Plan Work Assignments?

1. To select the right individual for the assignment

2. To anticipate and solve problems in advance

3. To meet schedules of time and cost

4. To minimize the waste of time resulting from delays in obtaining necessary information, services, and materials, and from inefficient job performances

The Methods of Planning Work Assignments

The method of planning work assignments for oneself or for others requires thought and organization. The following breakdown may appear unnecessarily involved, but it can be used to develop discipline in personal organization.

1. Group similar tasks.
2. Alternate difficult and simple tasks from time to time, if possible.
3. Do difficult or disliked tasks first, if possible.
4. Try to make all time productive; avoid interruptions.
5. Observe job performance to determine training needs.
6. Obtain the preliminary information necessary for planning any work assignment. What department policies might affect the particular assignment? Is the proposed job a function of another unit within the department? Are the necessary facilities available? Does the individual who will receive the job assignment have the skills and interests that the job requires? Can the proposed job be coordinated with work performed in other units?

Planning a Particular Work Assignment

Define the objective in writing. What is the expected result? Is this objective in line with unit and department functions? How does it fit into the overall scheme?

Analyze and list in writing the problems and facts. What do you know about the job? How does it compare with and differ from previous jobs assigned to the unit? When should the job be started and completed? Considering the entire work schedule of the unit, what decisions must be made? By whom? How would you do the job? What are the key points? Where are the pitfalls?

Who will solve the problems? What problems are the supervisor's responsibility? What problems will others solve? Who? What problems will the individual solve? The group?

What information is not available? Who will provide it? When will it be available? What should be done about it?

Who will do the job? The answer to this question depends on the job itself, the abilities and interests of the individuals in the group, their availability, and other factors. What are the qualifications of the individual selected for the job? What assistance will be needed by this person? How will it be provided?

Develop check lists for long jobs. Keep notes for checking progress of the job and for future similar assignments. Remember to schedule the progress and to report any new requirements for the job.

Making the Work Assignment to the Individual Selected

Before the supervisor makes a work assignment to an individual, he or she should realize that the interview should stress the positive rather than the negative aspects of the job. What does the subordinate need or want to know? How does the subordinate want or need to feel? What does the supervisor want to learn?

The following items are the major points to be covered in the work assignment, but the good supervisor must be aware of *how* he or she covers these points:

1. Make the job assignment at the appropriate time and place.

2. State the objectives of the job.

3. Indicate the effect of this job on the individual's other duties.

4. Discuss the problems the job entails and the importance of meeting schedules.

5. Relate the job to the individual's previous assignments.

6. Encourage the person to give opinions of the job and to offer suggestions.

7. Explain in detail the responsibilities of the job. To whom should the subordinate report? From whom can he or she get assistance? At what points should the subordinate make a progress report?

8. Inform the subordinate that any misunderstandings and questions should be resolved at this point. Ask the subordinate to summarize the job, and be sure that he or she has an adequate understanding of what is expected.

Follow-through

The following progress checks on the work assignment are required of the supervisor:

1. Review the progress of the job regularly.

2. Compile information that will be useful for similar jobs in the future.

3. Look for warning signals that indicate the subordinate is having

trouble. Is the job progressing according to the original time schedule? Are there frequent changes in the original plans? Is the subordinate complaining that he or she needs assistance?

4. If the job is not progressing well, the supervisor should consider changing the job objectives, discontinuing the job for a short time, giving the subordinate additional instructions, or supplying assistance.

DISCUSSION QUESTIONS

1. How can a supervisor improve the utilization of time and personnel?
2. Do the following work assignments constitute effective utilization of police officers? If so, why? If not, who else could and should do each of the following?
   a. Recruiting new police officers
   b. Enforcing parking meter violations
   c. Typing and filing index cards
   d. Operating an ambulance
   e. Classifying and searching fingerprints
   f. Dispatching mobile units by radio
   g. Photographing and fingerprinting prisoners
   h. Staffing the public information counter
   i. Investigating complaints against other police officers
3. What are some of the warning signals that something is wrong with a work assignment?
4. Why is the manner in which a supervisor presents a job assignment to a subordinate of such importance?

16 Performance Appraisal

The evaluation of a subordinate's performance is one of the most difficult and most important tasks of the police supervisor. The agency's training needs can be determined by performance appraisal because it identifies the weaknesses that limit an employee's effectiveness. It is invaluable in planning the distribution of the work force, because each individual's abilities are known. As a result of the performance appraisal, an employee can strengthen weak areas and may obtain a challenging job or a promotion, and the supervisor can utilize the work force more efficiently.

Types of Evaluation Systems

Performance rating systems are not extremely precise or scientific, but if they are administered properly they can provide relatively useful data. One of the main problems with most rating systems is the subjective, personal opinion of the rater. Because it involves human judgment, the process of evaluation cannot ever be completely objective.

Performance standards and measurement techniques have been developed and can be applied, the important aspects of performance can be emphasized and irrelevant details ignored, and common rating errors can be identified and eliminated. But, to date, no rating system has been devised that is really adequate, and there are nearly as many types of rating systems as there are agencies. The supervisor will have to modify any rating system chosen to meet the requirements of the agency. The three most commonly used rating systems are: the ranking system, the numerical scale, and the descriptive scale.

The Ranking System

The ranking system lists employees in order, according to their performance. The employee's overall performance is evaluated and ranked, or the work is divided into categories and the employee's skill in each area is appraised. A ranking system that works well for large groups is the *equal internal system,* in which the performance of each member of the work force is evaluated and, after the employees are ranked, they are divided into groups with equal skills. The most common ratings of the equal internal system are top, above middle, middle, below middle, and bottom. The employees in the top group are rated "excellent"; those placed in the above-middle group are rated "good"; those in the middle group, "satisfactory"; those in the below-middle group, "fair"; and those in the bottom group, "poor."

The Numerical Scale

The numerical scale places a numerical value on various characteristics, abilities, or aspects of the worker's performance. The total score is the rating.

For example:

Attitude toward others	1,2,3,4,5
Ability to learn	1,2,3,4,5
Dependability	1,2,3,4,5
Total score	15

Table 3.

In this example, the number "5" is the highest rating.

A minimum acceptable score for each category is generally established with the use of this rating system. Most numerical systems use an odd number of points so that the middle number is the average.

The Descriptive Scale

The descriptive scale is a qualitative, rather than quantitative, method of evaluating performance. The use of description allows a precision in ratings that is not practical with numerical scores. On a five-range scale, an employee might be rated for "Initiative" as: "unsatisfactory," "fair," "good," "excellent," or "superior." The words commonly used on a four-range scale are "unsatisfactory," "improvement needed," "competent," and "outstanding." The words commonly used on a three-range scale are "unacceptable," "improvement needed," and "acceptable."

A possible improvement of the simple descriptive scale is the use of phrases as indicators. For example, in rating an employee for "Cooperation," the rater selects the phrase from the list below that best describes the employee's ability to cooperate.

1. An obstructionist
2. Indifferent to others' needs and wishes
3. Cooperates much of the time; fairly willing
4. Works harmoniously with others; good team worker
5. Outstandingly cooperative; actively promotes harmony

Common Problems and Errors in Performance Appraisal

Frequently, performance appraisals are used to measure the wrong qualities. A supervisor often evaluates behavior rather than performance, and personality rather than accomplishments. These are the wrong measurements because the evaluation of an individual's personality and behavior does not necessarily contribute to the improvement of the individual's job performance.

Negligence and Incompetence of the Rater

A rater's negligence or incompetence may be one cause of inaccurate performance appraisals. Individual raters often apply different values to the various components of performance. Although this practice may never be eliminated, training on the mechanics of the rating system would help and should be required for each supervisor.

Prejudice of the Rater

All people have certain prejudices that affect their judgment and, though they may not realize it, their evaluations are influenced by a person's policies, religion, race, idiosyncrasies, and appearance.

If the supervisor is not careful, the fact that an employee is annoying can interfere with an objective evaluation of the subordinates's initiative, dependability, ability to learn, and so on. Such an appraisal is unfair to both the subordinate and the agency. One way to minimize this problem is to evaluate *all* subordinates on one aspect of job performance at a time.

Emphasis on Incidental Behavior

Most raters have a tendency to let one highly favorable trait color their judgment of all other traits—the exceptional trait becomes a "halo" that prevents proper consideration of the other factors. A highly unfavorable trait can also cloud a rater's mind, blocking out other qualities. It is essential to limit an evaluation to a specific time period; past performance must not influence the current appraisal.

Inadequate Knowledge of the Employee

The performance appraisal should be a continuous process. Unless the supervisor knows subordinates well, an evaluation of their performance cannot be of much value. The supervisor should arrange his or her schedule so that an adequate amount of time may be spent with each person to be evaluated. A knowledge of an employee's home life, interests, and problems can help the supervisor make a fair appraisal of the individual's performance. If there are new employees in the unit, the supervisor should not attempt to evaluate them until they are known well enough so that a valid judgment can be made.

The Error of Central Tendency

The error of central tendency is the practice of raters to avoid using the extremes of the rating scale. This tendency is especially common when raters have inadequate information and would be unable to defend a high or low rating. In effect, they reduce the precision of the rating system to conform to their own ability to rate.

Leniency

Some supervisors overrate their subordinates to avoid antagonizing them, particularly those with strong personalities. Or supervisors make the ratings support their recommendations—low ratings reflect unfavorably on the efficiency of the supervisor in selecting, training, and disciplining subordinates. The supervisor may also overrate an employee who is older and has had more experience in the police service than the supervisor.

Few employees are either good or poor in *everything*—most of them rate high in some categories, average in others, and low in still others. If ratings are too high, they are not fair to other employees or to the department.

Severity

Sometimes supervisors are too severe in their appraisals of subordinates. For personal reasons, they find it extremely hard to give a high rating. Obviously, ratings that are too unfavorable are just as unfair as those that are too favorable.

Appraisal of Potential Value

A subordinate should be rated on actual job performance and *not* on what he or she may do or may become in the future. The appraisal of current performance is far more accurate than that of potential performance.

Any appraisal of an employee's potential value should be made after a performance rating has been completed and all other information about the individual has been gathered and studied. The two types of appraisal should not be confused.

Measurement Considerations

It is not possible for supervisors to *consistently* define and measure something as intangible as personality traits; this is a job for a psychiatrist or a psychologist. While all human judgment is distorted to some degree by prejudice, it is much more pronounced when evaluating personality traits. Prejudice may work for or against the person being rated, but either way it reduces the accuracy of the observation.

There is little agreement about what personality traits are required in order to be a successful officer. It is also extremely difficult to set up a rating form that consistently indicates the extent to which the person being rated possesses certain traits. A rating form is useless unless various raters can agree on the exact definitions of the factors they are rating. Raters have a tendency to apply their own definitions to abstract terms such as drive, initiative, and judgment.

In effect, some rating forms ask the supervisor to measure what the subordinate *is* instead of what the subordinate *does*—to observe the *employee* rather than the employee's activities. An effective rating system requires the supervisor to deal with facts and to stay away from abstractions. This can be accomplished only if the rater has observed and inspected the performance of the officer to be rated in all phases, and if observations have been documented on a daily basis. Too often, incident reports are made only when there is an infraction of the regulations or when an outstanding job has been done. Then, when the rating period arrives, the rater bases an "opinion" of performance primarily on that information and whatever else he or she can recall. The rating should be a measurement of daily performance, not the opinion or judgment of the rater based on isolated reports.

Abstract qualities such as initiative, attitude toward fellow officers, judgment, and loyalty can be rated, but only if they are related to specific aspects of the job. A low rating on "judgment" must be justified on the basis of specific examples of how the individual exercised poor judgment, not on the basis of vague generalities. When attempting to rate an abstract quality, the supervisor must evaluate to what degree the individual has specifically demonstrated his or her abilities, or lack thereof, in regard to the trait being considered.

A useful device for the supervisor is a record of impressions written while working with subordinates. A daily record of good and bad performances of subordinates will provide a guide for the supervisor when later evaluating subordinates. The supervisor should not attempt to rely entirely upon memory, because it is impossible to remember all the pertinent details about each person's performance. On the other hand, the supervisor should take care not to become known to subordinates as a "snoopervisor"; notes should not be made in the presence of subordinates.

A supervisor who is keenly interested in improving an individual's performance should conduct a counseling session with the employee prior to the submission of a formal rating. This conference will provide an opportunity to discuss any deficiency that might be cited in the performance evaluation and will give a conscientious employee a chance to modify personal behavior so that an unfavorable report will not be necessary. In the event that the behavior does not change, a comment in the formal report alluding to the counseling session or sessions is of great benefit to the supervisor. The employee would not be able to say that he or she had not been advised of any performance deficiencies.

Elements of a Good Performance Appraisal System

Discussions between supervisors and subordinates should stress the point that performance appraisals are based on job performance rather than on the value judgments of the supervisor. Performance standards often are established by a mutual understanding between the supervisor and the subordinate. Therefore, supervisors must demonstrate skill in conducting performance reviews and counseling interviews, as goals are often discussed during this process. See Chapter 12 for a discussion of the techniques of counseling and interviewing.

Subordinates have a right to know what their supervisor expects of them. Their jobs and standards of performance must be well defined. The evaluation of their performance should be based on a combination of the following: quality and accuracy of work; quantity of work; manner and methods of doing the work; length of time required to perform specific tasks; adherence to deadlines and work schedules; end results of the job in terms of case closure rates, crime reduction, enforcement indices, personnel complaints and commendations, success of prosecutions, etc.

For the convenience of the rater, personnel records containing past performance appraisals, job progress reports, and special training and education should be available for examination. These might disclose a change in performance, either good or bad. Performance changes often go unidentified in large organizations.

Supervisors should be properly trained in the fundamental principles of rating and must be thoroughly familiar with the system and whatever form is being used.

Differences in values must be resolved, and supervisors must be aware of common errors and pitfalls. Skills must be developed for the performance review and counseling interview. Follow-up evaluations by management must be made to assure that the purpose of the rating program is being achieved.

How to Improve Performance Appraisals

Supervisors can improve their appraisals of subordinates by utilizing the following practices:

1. Study your subordinates. Observe them at work and talk to them at frequent intervals. Listen to what they have to say, and record comments and behavior that have a bearing on their job performance.

2. Before beginning an appraisal, cross off the name of any person you do not know well enough to rate.

3. Study your department's rating forms carefully, especially the definitions for each factor.

4. Read the rating form instructions carefully.

5. Avoid giving too much weight to isolated instances.

6. Do not allow personal feelings and prejudices to govern your appraisal.

7. Do not overrate or underrate, but strive for a fair appraisal. Remember the time period the rating covers.

8. Appraise subordinates on past and current performance and behavior, not on their potential.

9. Appraise all subordinates on one trait before beginning a second.

10. *Remember that rating is a mental process* that requires more than filling out a form.

The supervisor must not overlook the obligation to recognize good work as well as to call attention to areas that the employee needs to improve. It takes real skill to do this properly. The supervisor must observe continuously to be able to rate, and must record the results of observations and discuss these results with the employee. This chapter has been dedicated to calling attention to the many basic concepts, errors, and common problems in rating work performance.

DISCUSSION QUESTIONS

1. What are the advantages to the supervisor of the work performance review and discussion of it with the employee?
2. What are the advantages of the work performance review and its discussion to the employee?
3. What aspects of a supervisor's job are affected by the work performance review and discussion of it with the employee?
4. How would the supervisor's job be made easier by eliminating performance reviews and discussions?
5. What is the added value of periodic work performance reviews in addition to everyday discussions of work performance?

17. Women and the Supervisory Role

In recent years, events have heightened the American woman's perception of herself and her role in society. Today, more and more highly educated and capable women are making the law enforcement profession their career. Protected by law against discrimination in selection or deployment, and no longer relegated by tradition to less sophisticated semisworn positions (such as matron, juvenile officer, meter maid, etc.), a growing number of women are now serving their communities as sworn police officers in all capacities, including patrol. In addition, many are being advanced into higher supervisory rank as rapidly as their abilities, the available openings, and civil service regulations permit.

The potential gain for law enforcement is tremendous. The traditional selection of patrol officers on the basis of male gender, strength, and size will soon be universally replaced by more valid criteria. The more than one thousand female police officers currently assigned to patrol attest to selection criteria more closely matched to the actual job requirements. Furthermore, these female officers are proving that women, properly selected and properly trained, can assume the responsibility for any position within the law enforcement profession.

Where Were Women Serving?

For decades in this country, women have served in a support capacity to an all-male police force. Often these women were paid far less for doing the same work, and it was only a matter of time before the question of equality in position and pay would arise.

Given a new consciousness by the strong and enduring social force known as the Women's Movement in the late 1960s, female applicants for police department positions grew rapidly in number. But police executives in most cities, guided by tradition, fear of the unknown, or a combination of both, refused to allow females to assume patrol duties. In March, 1972, however, the United States Congress passed the Equal Rights Amendment to the Constitution, and the barriers began to fall for women in all professions.

However, law enforcement remained rigid against the pressure and cited the aforementioned criteria of size and strength as prerequisites for individuals to work patrol. Despite this attitude, the number of women occupying prominent roles in police agencies as working patrol officers and the number of women occupying supervisory positions have quietly increased throughout the country.

Where Are They Serving Now?

According to the 1974 Police Foundation report entitled *Women in Policing: A Manual,* "in 1971 there were fewer than a dozen policewomen on patrol in the U.S.; in 1974 there were close to 1,000. In 1971, there were only a few women in police supervisory positions; in 1974 there were several hundred women sergeants, lieutenants and captains supervising male and female patrol officers and detectives."[1]

Clearly, the selection of individuals for patrol duties and supervisory duties is being based on performance, capability, and merit, regardless of sex. What, then, are some of the special problems facing women police officers and their male partners, women serving in supervisory capacities, men supervising women, and departments which have only recently begun recruiting women officers?

Enter the Female Police Officer

"It'll never happen in this department," many veteran policemen have regrettably uttered. They have often been forced against their wills to accept the reality of female officers. What does this resistance mean to the newly trained female rookie? To her, it's confusing, frustrating, and often a great impediment to proper field training and orientation to the requirements of the job. Men who do not accept women in patrol work present intense problems for the male supervisor, and it is imperative that he place any of his negative feelings aside and lead the resisting men to an acceptance of the women and their abilities. The supervisor must encourage humor, but must not allow derogatory remarks by male subordinates about females, for this only perpetuates the negative stereotype of the incapable female. Rather, the supervisor should involve objecting males in the preparation of training programs and other facets of the orientation and indoctrination of females.

With this type of positive supervisory leadership and the resultant positive atmosphere created by the supervisor, traditionally closed attitudes among subordinates will begin to open and change. And as the female officer "proves" herself in carrying out the patrol duties (as all rookies must do), behavior toward female officers will improve noticeably. The supervisor must create an organizational environment in which negative attitudes toward women will not be tolerated. Instead, a positive program must be developed by each supervisor to eliminate isolation of women officers and encourage their acceptance into all phases of police work.

The female police supervisor probably has experienced several years of attitudinal and behavioral problems in achieving her promotions, but regardless of her own struggle in winning peer support, she should remember that new recruits, both male and female, will probably be working under a female supervisor for the first time and could experience attitudinal problems of their own.

Recruitment

In many agencies, the recruitment of new personnel is a function of a separate division within the organization, but the supervisor of any unit who would strengthen his or her department must continually be alert for individuals who would make good police officers. With the entrance of women into the profession, agencies now have reassessed the factors which make a good officer. These factors are included in the list below.

Future recruitment programs must be developed to identify persons who:

1. Are able to accept responsibility for the consequences of their own actions

2. Can understand and apply legal concepts in specific situations

3. Can clearly handle challenges to their own self-respect and status

4. Are able to tolerate ambiguous situations

5. Exhibit willingness and desire to participate as team members

6. Display skills associated with group interaction or skills facilitating team building

7. Have the physical capacity to subdue or control persons

Agencies should recruit personnel capable of developing an inner sense of competence and self-assurance so that under conditions of stress, conflict, and uncertainty the individual officers will be capable of responding flexibly and in a relatively dispassionate manner rather than rigidly, emotionally, or defensively.

Training

Traditionally, the job of the police officer is one of crime control, but increasingly it is also one which is deeply involved with the provision of social services to the community. It is incumbent upon the training supervisor to remember that the heart of policing consists of working with human emotional problems—often at crisis level. The increased number of female police officers, who are generally smaller in stature and weaker in strength than men, suggests that skills *other* than size and strength may be most important to the effective delivery of police services.

The American Bar Association notes that, "it may be that skill in interpersonal relationships may be as important as physical fitness in equipping the officer to cope with potentially dangerous situations."[2]

The greatest reservations about the assignment of women to patrol appear to come from doubts about their physical ability. The training officer would do well to note that as more and more departments develop job-related standards for training, it appears that leverage strength, rather than brute strength, is most important. Training of women in the use of martial arts, with a program of continual retraining, will promote self-confidence in the women and respect from their male counterparts. Regardless of the type of training given, it is of primary importance that all officers take the *same* training with the *same* standards for passing. All tests must be job-related and designed so as not to discriminate unfairly.

Evaluation of Performance

We have previously mentioned that the question of whether or not women should be allowed to perform in patrol capacity has never been based on material fact, but rather on tradition, superstition, and male insecurity. With data coming in supporting the ability of many females to perform as well as males in patrol functions, the question of gender in evaluation becomes irrelevant and is replaced by the question, "What kind of person makes a good police officer?"

To the supervisor dealing with female recruits for the first time, the responsibility of subordinate evaluation is especially acute, and extra care must be used to eliminate factors which may bias either the information which the supervisor is receiving concerning the woman's performance, or the supervisor's report itself. Personal attitude, as discussed previously, may definitely alter the legitimacy of the report either against or for the officer, just as a supervisor's fear of contradicting what would appear to be new departmental policy or practice would hamper judgment.

In any event, the special attention given by the dedicated supervisor to this task will insure the fairness of his or her personnel evaluations and may help to clarify the kinds of performance and personal characteristics that make for a good police officer. This comprehensive monitoring of individuals with respect to quality of performance against measurable and equal standards will guide the rating supervisor and the department administration in the elimination of individuals unfit to adequately carry out the police function.

Promotion and Opportunity

It follows from the above discussion that promotion for both men and women must be based upon competence and ability. But indications are that not all officers receive equal opportunity for the development of these promotional characteristics. It is up to the supervisor to ensure that all female officers receive the opportunity to perform. If women are to be evaluated for promotion to patrol supervisor, they must have considerable experience in patrol work, and during this period they must have a close and constant interaction with their supervisor. It should be noted that many women may be initially reluctant to perform full patrol duties. This should not be misinterpreted as lack of ability, but as lack of experience. The supervisor should encourage the officer to expand her abilities gradually but consistently, realizing that female officers may be less fully acquainted with the range of patrol responsibilities than their male counterparts. If positive encouragement fails to bring about the required confidence and the female officer is transferred out of patrol as a result, then it is the responsibility of the supervisor to relate this information to the selection and training staff in order to improve officer selection and prevent future transfers. This procedure would apply equally to male subordinates. As long as all promotion lists, job assignments, training, evaluation, rewards, and discipline are equal for all officers, male and female, the department which promotes women to supervisory and administrative positions can be assured of the fairness and the quality of its promotions.

Special Considerations

As the number of women in patrol and supervisory positions increases, initially perplexing internal problems such as locker room assignments and uniform design will take on much less importance, and external concerns will begin to surface. It has been said that people are often afraid of that which they do not understand, and so it is with women entering the traditionally male-dominated police profession. Both male and female officers' spouses and families can be expected to react with concern to the assignment of male and female partners. The supervisor has an opportunity here to educate people outside the department. It is vitally important that the officer's spouse be provided the opportunity to ride along and observe the patrol function, to participate in rap sessions—which should be frequently scheduled—and to get involved in any auxiliary programs offered by the department. In this manner, greater information can be transmitted and the initially negative response of most spouses can be turned into a positive support for the integrated patrol program.

Problems may develop from two additional areas. Supervisors should be aware of the possibility of an initial public response which might be less than supportive. Most community leaders are still men, and many hold traditional attitudes. A positive public awareness program which does not place undue pressure upon any individual female officer might be a solution. Or, as has been the case in many communities, it might be advantageous for little or no public mention to be made of the introduction of women to patrol. In any event, supervisors should be aware of public interest in this subject.

Finally, the female supervisor can expect to encounter an inordinate amount of male resistance initially and should be prepared to deal with male peers or subordinates who are seriously threatened by the entrance of women into the traditionally masculine system. Dealing with these feelings rationally, openly, and fairly will greatly increase department morale and improve the supervisor's effectiveness of command.

The difficulty encountered by women in gaining acceptance in the law enforcement profession is not new. Many other minorities in the past have encountered such difficulties. The uniqueness of female officers and police supervisors will soon wear away, and when the newness disappears, it is our belief that the problems associated with this development will also disappear. When women prove both to themselves and to their male colleagues that they can be able and competent police officers and supervisors—a credit to their organization—they will be accepted fully as part of the modern police force.

DISCUSSION QUESTIONS

1. What was the traditional role of women in the police force? How has this role changed in recent years?
2. What are some of the problems facing women in the police force today? What special problems do women supervisors have?

3. What problems are supervisors with female subordinates faced with, and how can these problems be solved?
4. What traits should police agencies look for today as they recruit new officers?
5. What kind of training should be given female police officers?
6. What kinds of public reactions can be expected to the increasing number of female officers, and how should these reactions be dealt with?

NOTES

1. Catherine Higgs Milton, *Women in Policing: A Manual* (Washington, D.C.: The Police Foundation, 1974).
2. American Bar Association Project on Standards for Criminal Justice, *The Urban Police Function* (New York: American Bar Association, 1973), pp. 207-208.

18 The Supervisory Training Function

We are in an era in which increasing importance is being placed on police training. The physical face of America is being transformed. There are new people and new communities, and old communities completely change their character in less than a generation. The challenge to the police force is more than to keep up with the growth; the stresses and dislocations magnify police problems far beyond the dimensions of the growth itself. The demands on the police are not only greater, but are also new and different. The professional police officer must be concerned with many matters—personnel, records, accounting, detention facilities, services, and public relations. Complicated federal, state, and local laws that deal with labor and minority groups have been enacted. All these factors plus the difficulty in recruiting and the turnover of personnel demand more intensive training than ever of police officers by supervisors and administrators.

The establishment of statewide training standards in New York and California has intensified national interest in police training. As of early 1976, forty-five states had passed legislation that created a Commission on Peace Officers' Standards and Training (POST) and three states had such legislation pending. It is probable that some variation of a POST commission will be created ultimately in each of the states. Police agencies everywhere are showing an interest in achieving complete professionalism in the police service. This goal can be achieved only through pre-service and in-service education and training of *all* police personnel.

There has been tremendous impetus in police training in California since 1959, when legislation created the Commission on Peace Officers' Standards and Training. Basically, the Commission has the power to establish minimum standards of recruitment and training for peace officers in those communities which choose, by passing local legislation, to participate in the Commission's program. In return for maintaining the standards set by the Commission, the community is reimbursed by the state for 50 percent of the cost of the training program. The money for this training fund comes from an assessment levied against all of the fines paid by individuals for criminal offenses. Thus, in California the program operates without cost to the taxpayer. More than 95 percent of the law enforcement officers in California are now trained under the POST program. During the first years of the program, funds were only sufficient to meet the needs of recruit and supervisory training. However, POST now subsidizes training of middle management, technical, and administrative personnel.

The New York standards and training program is the Municipal Police Training Council, in which is included the Office for Local Government of the Executive Department of the State. This council operates in a manner very similar to California's POST. The program in New York includes all of the police departments within the state with the exception of New York City, which is the largest police organization in the country, with more than 35,000 personnel. The city of New York's police department was one of the pioneers in police academy training, and apparently, when the Municipal Police Training Council was established, there was no need to include New York City within the scope of the council's jurisdiction.

An increasing number of police agencies are requiring that applicants for recruit positions have completed at least two years of college education, and some now require that the applicant hold a bachelor's degree. Those agencies that have adopted the college requirement have found that the recruitment cost has dropped because they have to screen fewer applicants for existing vacancies.

It is important to realize that the words "education" and "training" are not synonymous; too often, they are used loosely and without proper definition. When we speak of education we mean learning for the learner's sake—the opportunity for self-improvement in the learner's chosen area. Education often deals with fundamental concepts and ideas rather than with specific topics such as attitudes, knowledge, understanding, and insight.

Training is defined as learning for the job's sake. Training places the emphasis on skills and is related to the job. Training should help improve competence by giving one knowledge that can be applied in daily work. Pre-service training prepares one to engage in a specific occupation, while in-service training is designed to improve performance or to prepare an individual for a specific assignment. Thus, we can see that college work may often result in both education and training, depending on the course of study.

Since World War II, the growth of criminal justice and police education curricula in colleges and universities has been spectacular. In 1945 there were only a few such programs scattered around the United States. Today, however, particularly since the inception of federal support to criminal justice personnel through the Law Enforcement Assistance Administration, the number of college and university programs has jumped to well over three thousand.

The pioneers in police education programs are well known: the School of Criminology at the University of California at Berkeley, now defunct, Northwestern University's renowned Traffic Institute, the University of Southern California's Delinquency Control Institute, Michigan State University's National Institute on Police and Community Relations, and programs at Indiana University and the State Universities of Florida and Washington. In New York City, the John Jay College of Criminal Justice, with more than six thousand students, has been created within the City University of New York.

In California alone there are about fourteen four-year colleges and universities and over ninety two-year community colleges presently offering professional criminal justice and police education. California State University in Sacramento, one of the leaders in the field, has expanded its curriculum offerings to areas of the state where four-year degrees were previously not available. This has been done through the university's external degree programs, which are presently operating at six locations, some as much as 350 miles away from the main campus. New college programs will continue to be created throughout the country as progressive police administrators demand a higher level of education of both recruits and in-service officers. Many agencies have ongoing educational incentive plans which enable officers to upgrade their salary level by varying percentages as they achieve college degrees.

As a key person in the education and training of personnel, the police supervisor must not only participate in the training of departmental employees, but must also be thoroughly familiar with what education and training are available outside the department so that he or she may properly function as a training counselor. The supervisor must know how to determine training needs as well as how to impart the needed instruction. Each of these functions will be described in the ensuing discussion.

The Supervisor's Responsibilities

Determining an agency's training needs is a responsibility of top executives. "Training Director" is usually the staff title in the large agency while the "Chief" assumes responsibilities for training in the small agency.

The supervisor's responsibility for instruction rests in the following areas:

Administration

Planning corrections and changes in the training program

Training

1. Assisting in the development of the training program through a thorough knowledge of teaching methods and sources of training

2. Ensuring proper induction training of new appointees

3. Acting as counselor to older officers for promotion studies

4. Being a good instructor, able to teach new employees the "know-how," and seeing that they learn through each reprimand, grievance, or new situation

Supervision

1. Checking the results

2. Troubleshooting

Determination of Training Needs

A number of methods may be used to determine what training is needed:

1. Conferences with supervisors
2. Surveys
3. Questionnaires and tests
4. Major changes in the activities of the department
5. Investigation and conferences with supervisors in other departments
6. Input from the community

Instruction must be properly planned to be of any value. Too frequently, police agencies develop a training program without giving thought to the following important criteria for development of planned programs:

1. For whom is the instruction intended?

2. Does the program meet the needs of the department?

3. Does it meet the needs of the personnel for whom the training is intended, considering their age, experience, and assignments?

4. Does it meet the needs of the municipality or locality?

5. Is it a temporary shot-in-the-arm substitute, or is it a well organized and continuous program?

There are a number of phases of training in the police service. Any of the following ones may be part of an individual department's program:

1. Pre-service (colleges or junior colleges)
2. Cadet or recruit training
3. In-service training (intermediate or advanced)
4. Specialized or technical training
5. Administrative-supervisory training

Essentials of Efficient Training Programs

Those who have had experience in the organization and the operation of training programs know that, "successful programs, those that are accomplishing the purposes they were set up for, possess certain characteristics, regardless of the specific occupations for which training is being given."[1] Where no organized training program has been developed, it is advisable to contact and seek advice from administrators of successful training programs to avoid some of the common mistakes. Good sources of information on such programs are the Commission on Peace Officers' Standards and Training, the training officers of other law enforcement agencies, and local school authorities.

Four Principal Characteristics

Four principal characteristics are recognized as essential to an efficient training program.[2] These are: selection, competence, functional subject matter, and good working conditions.

Selection   It is obvious that a program of professional training for law enforcement is designed for either experienced officers or prospective, or inexperienced, officers. For the greatest efficiency, officers who have had experience in the service should be trained separately, not with relatively inexperienced officers or prospective personnel. The difficulties encountered by an instructor who attempts to handle any unit of training with a heterogeneous group are obvious. The effectiveness of the training is sharply reduced by attempting to handle a single group composed of personnel with widely varying backgrounds, for whom the training objectives are quite different.

Competence   The instructor should be competent, not only from the standpoint of knowledge and skill regarding the practical work of law enforcement but also from the standpoint of ability to teach what he or she knows. When it appears appropriate, there should be no hesitance about recruiting instructors from outside the department, including civilians who possess expertise that is needed in specialized fields. A good example of this would be the use of a psychologist to teach certain aspects of human relations and related areas.

Functional Subject Matter   The course of study should be based upon what is commonly referred to as "functional subject matter." Law enforcement courses must not be formulated on the basis of hunches, imagination, or what someone else may have done independently, with no consideration of the task to be performed. From the standpoint of professional education and training, these unscientific methods and approaches to the training problem are so ineffective that they merit no serious consideration. What the officer does and what this person needs to know to be a successful worker in law enforcement are the keys to instructional content. The instructional objective should be to make available to the learner the sum total of all the skills, abilities, concepts, and knowledge that he or she needs to develop into a successful officer.

Working Conditions   Working conditions should be favorable and conducive to a good training environment. Working conditions should include:

1. Training groups of a manageable size
2. A sufficient amount of time allocated for training
3. Good physical facilities and equipment

Orientation of the Recruit

Before a police recruit can perform the simplest task, he or she must have adequate knowledge. Police work differs so much from almost all other fields that the average recruit has no background or experience even remotely comparable to it. When police recruits complete the entrance examination and qualify for positions on the force, they are eager to complete their training, take their places on the street, and leave behind as quickly as possible their rookie days. How, then, are they to receive further training so that they can qualify as full-fledged police officers?

Reception by Veteran Officers

The veteran members of the force should be prepared to receive recruits kindly. They should be instructed to give recruits every assistance and to refrain from wisecracks and undue criticism lest the recruits feel that they are not wanted on the force. The average police officer is slow to accept rookies as equals. Until they have proven themselves, perhaps a certain amount of this reserve is justified. But under no circumstances are recruits to be belittled or made to appear ridiculous before the public or other members of the force. If recruits make mistakes, they should be corrected in private. Recruits can retain the self-confidence so necessary in police work if they are corrected privately; public exhibitions destroy that confidence.[3]

Recruit School

Much of the training the recruit must master is dry and uninteresting. The learning of laws, ordinances, rules and regulations, and rules of evidence becomes a boring task if it is not presented properly. Too many recruit schools start their beginners with long hours in the classroom, subject them to uninteresting lectures, and force them to memorize long passages of written material which are quickly forgotten once a passing grade has been obtained. The method of presentation should be diversified with motion pictures, field trips, demonstrations, and discussions. Student participation in all these learning activities should be encouraged. Instructors should use examples and personal experiences to make their lectures interesting. The school day should be broken up so that it is not all lecture. Physical training, gunnery, defensive tactics, and patrol tours of a district in the company of an older officer, etc., can be utilized to lend more interest to training and to make it more meaningful.

When the recruit is trained by accompanying a police officer on a tour of duty, it is extremely important that the new recruit be placed with an experienced officer who has a professional attitude and approach to the job. The choice of the wrong senior officer can undermine the entire training program.

The Line Supervisor's Responsibility

In many agencies, the real burden of responsibility for field instruction falls upon the line supervisors. Some administrators consider this to be one of the most important of all the supervisory functions. It is the supervisor who actually carries out the field instruction or sees that it is done properly by senior officers.

Field Instruction

Many police agencies—for example, in Wichita, Berkeley, Culver City, and Pasadena—issue manuals to field instructors which remind them of their obligation to the recruit and suggest outlines for the field training courses. In patrol work, the instructor must be able to explain how to handle any situation that occurs in the field. As opportunities arise, the instructor must perform tasks in the presence of the recruit. Finally, the recruit must perform the same or similar tasks. This method has produced excellent results and is highly recommended because it provides the field instructor with a guide for training the recruit and also provides a record of the recruit's progress. One of the best manuals for field instructors seen by the authors to date is the *Field Training Guide* published by and available from the California Commission on Peace Officer Standards and Training. The guide is 182 pages in length and serves as a model from which any police agency can develop a manual suitable to its own needs.

In the agencies which issue manuals, field instructors frequently attend conferences during the training period to discuss the progress of the recruits and to receive information on methods of instruction. The average recruit will enjoy the field training phase more than the classroom training. The use of actual field experience helps produce a well-rounded training program, a happy result to which too much attention cannot be directed. It is during the first days and weeks in the department that the recruit forms attitudes and opinions of the police service which may have a lasting influence, perhaps for the recruit's entire police career.

The Induction Interview

In addition to being responsible for training, the supervisor conducts the extremely important induction interview. In this interview, the supervisor is a builder of attitudes. He or she has the chance to focus, or integrate, as well as explain all of the material that the new officer has received. When the supervisor greets the new officer, a reference to their first meeting or a friendly continuation of some topic previously discussed helps put the officer at ease. A supervisor may schedule the interview for a coffee period to make it easy and pleasant. One excellent supervisor allows up to an hour for this interview, feeling that this time is well spent. The officers appreciate this thoughtfulness, remember it with pleasure, and reward it with hard work.

In order to create enthusiasm the supervisor must know the job, believe that it is worthwhile, and be able to communicate this knowledge and feeling to the new officers. The induction interview gives the supervisor a good opportunity to practice what he or she has learned in the art of communication and to turn learning into *student* know-how. The supervisor should remember that during induction the new officer is apt to be far more eager to listen and learn than he or she will be later. The new officer's eagerness and zeal should be encouraged and used by the supervisor in every possible productive way. It may be the only time that every element of the job is a positive one.

The first phase of the induction interview consists of an outline of the work of the department and the names of the other divisions and supervisors. It may be wise to give the new officer a general idea of the background and experience of the officers in supervisory positions. This indicates that there are opportunities for promotion and builds ambition and hope.

The next step should be an explanation of the department's rules and procedures. If the supervisor has had a hand in formulating these rules and feels they are satisfactory, he or she will naturally give them enthusiastic support. This is no time for a supervisor to recall any past difficulties in applying the rules or to express feelings about any people with whom he or she has had disagreements over department procedures. The officer will be more impressed by the supervisor's behavior and attitudes than by what he or she says.

The important person in the situation is the appointee, not the supervisor. Clearing up any questions the new officer may have is important. Crucial points that may come up in later discussions or that may be a source of future disagreement should be reemphasized, even at the risk of repetition. First impressions are most important. The supervisor must make the appointee's first day on the job a positive one. Another supervisory obligation is to remind older personnel of their responsibilities to the new officer.[4]

Planning the Recruit Training Course

In planning a recruit training course, care must be exercised in the selection of subjects and in specifying the emphasis to be placed on each one. If the program is not carefully planned, relatively unimportant subjects are likely to be over-emphasized and important ones forgotten. As every police administrator knows, police work includes such a multiplicity of tasks that, if each were to be thoroughly covered before the recruit left the classroom, he or she could spend an entire career in the recruit school. In those states that have a Commission on Peace Officers' Standards and Training, a core curriculum for recruit schools usually is specified by the commission. However, no matter which course is the basis for the recruit school, the curriculum must be based on an analysis of the police officer's job.

A long overdue innovation in recruit training has appeared within the past several years. More and more police agencies are including orientation for spouses, including ridealongs, as an integral part of recruit training. Many times this phase of recruit training is conducted by a qualified psychologist who will delve into the problems confronting families of police officers who are subjected to extremes of stress in their daily work routine. It is indeed a serious oversight that in an occupational group with one of the highest rates of divorce, alcoholism, and suicide of any group in the United States today, so little attention has been given to the problem of stress and its effects on the individual officer and his or her family.

Basic Job Information

Every new employee has a right to know those things about the organization that will affect him or her directly. This information reduces mistakes and confusion and helps the employee to get off to the right start. The following list covers most of these items:

1. Salary and deductions
2. Duties and responsibilities
3. Working hours, shifts, pay, benefits, sick leave, etc.
4. Rules, regulations, and policies
5. Employee's relationship to the department and to the public
6. Proper conduct
7. Possibilities and advantages of the job
8. Employee's place in the organization—the person to whom he or she reports
9. Chain of command
10. General functions of other divisions and units
11. Job security and protection
12. Name of the collective bargaining agent

Basic Work Information

The new officer also must know certain things before he or she can work alone. It is generally considered that training in all subjects listed below would be an ideal rather than a practical course of instruction. However, the given list is a comprehensive training plan, and probably includes most of the information a new police officer should receive. Instruction should follow this sequence as closely as possible, as the easier tasks are listed first. A training program can give more instruction than a new officer can absorb at one time. The quantity of material to be covered should be left to the judgment of the instructor, who must remember that if the student doesn't learn, the instructor hasn't taught. Before the recruit can work alone, he or she must have information about each of the following items:

1. Geography of the city and county
2. Referral agencies—hospitals, relief agencies, other police departments, etc.
3. City and county governments, their departments and officials
4. Care and operation of a police car and other equipment
5. Field communications, including call boxes, mobile radios, and allied equipment

6. Equipment that is necessary for the officer's use
7. Use of firearms
8. Beat boundaries and beat patrol procedures
9. Vehicle, criminal, civil, and local ordinances
10. Weaponless control and self-defense
11. Laws of arrest, search, and seizure
12. Mechanics of arrest
13. Keeping a notebook
14. Taking and making all reports for the job
15. Handling and preservation of evidence
16. First aid
17. Traffic activities
18. Jail procedure and booking
19. Crime prevention in general
20. Court demeanor
21. Basic knowledge of public relations
22. Community and racial relations
23. Civil defense procedures
24. Mob and riot control

The reader could no doubt add to this list, but each of these topics should be considered in its broadest context.

Advanced Officer Training and Education

In advanced police training, which is designed to improve the police officer's efficiency, the objective is the systematic improvement and upgrading of those officers who have completed their basic training and probationary period. The courses are designed to acquaint them with new techniques and new regulations. It is at this point in an officer's career that the supervisor can either increase or decrease the officer's confidence in him- or herself, the supervisor, and the department.

Motivation for More Knowledge

The supervisor must motivate subordinates to seek advanced training and education, and must be able to advise them on the continuation of their training beyond that offered by the agency. Certain factors presumed to create an interest in education are: a clear-cut definition of goals and objectives, the supervisor's confidence in the officer's ability to do the job, the officer's desire for recognition, and human curiosity. Fear should never be used as an incentive to promote training. An increasing number of police agencies are using incentive plans to persuade their personnel to seek advanced training and education. Pay increases based on the number of college units completed is one of the most effective incentives. Another incentive is the requirement of advanced education for promotion.

Roll Call Training

One of the most common programs for advanced training is some variation of the roll call training that was pioneered by the Los Angeles, California, Police Department. The most usual pattern is a fifteen-minute training session given by the supervisor each day during the roll-call period.

Training Bulletins

Some agencies have developed their own series of Roll-Call Training Bulletins. For those that do not have the resources for this, the International Association of Chiefs of Police has developed several series for general distribution, including one called *The Training Keys* and another called *Sight and Sound*. They are available for a nominal fee. After discussion of a topic, a bulletin that covers it is distributed. Its purpose is to reinforce what the officer has learned. This training program has proved very successful in those agencies that have adopted it. It is an extremely flexible training medium because bulletins can be written to cover topical problems and situations.

Periodic Courses

An additional source of advanced training is the periodic course, conducted by the agency, that covers special subject matter. For smaller agencies in a reasonably small geographic area, a combined operation in the form of a zone school can be initiated. Instruction in almost any police subject can also be obtained by local police agencies from the Federal Bureau of Investigation. Departments contemplating the establishment of schools of any type should confer with the officials of their local school district to find out whether they can coordinate the program with a local high school or junior college district and thus have it become eligible for funds from a federal trade and industrial education department. If the state has a Standards and Training Commission, all training programs should be approved by the Commission, particularly when partial reimbursement is available for those schools that are approved.

College-Level Programs

It is essential that the supervisor be thoroughly familiar with any college-level programs that are available in the area. There are now more than three thousand colleges and universities that offer some form of police education or training, ranging from highly theoretical courses to the most practical type of field training. Periodically, the International Association of Chiefs of Police publishes a volume listing these institutions. *The Police Chief,* the monthly publication of the IACP, includes a Training Calendar that lists the types of training being offered, the dates and locations, application deadlines, costs, and hours of instruction. The Training Calendar is one of the best sources of information on available training.

Some of the best known training and education programs that are available regularly are the following:

> 1. The Delinquency Control Institute at the University of Southern California, which is a specialized training program designed to prepare law enforcement officers and others in related fields to work most effectively with young people who have problems. The program runs for twelve weeks and is offered twice a year.
>
> 2. The Traffic Institute of Northwestern University in Evanston, Illinois, offers an extensive training program ranging from some unit courses, conferences, and seminars of various lengths to the nine-month traffic-police administration training program.
>
> 3. The FBI's National Police Academy presently holds four eleven-week sessions each year. The admissions standards for the Academy are very rigid, and the men and women chosen for this training are expected to return to their respective departments and teach what they have learned to others in the department.
>
> 4. The California State Universities at Los Angeles, Long Beach, San Jose and Sacramento periodically run special institutes for police officers in subjects ranging from polygraph training to middle-management training as well as many other police-oriented subjects.

There are too many college-level programs available to list all of them here. Much depends upon the location of the training program and the availability of funds to send men and women to specialized schools.

In many states, the highway patrol and other state police organizations conduct academy and specialized training programs which members of other departments may attend. Also, the training academies of many of the larger metropolitan police departments and large sheriff's offices encourage smaller departments in the surrounding area to send their personnel for various types of training. It is not at all unusual to find a large city training academy supplying all of the training for many small communities in the general area. A core program is offered to all personnel at the academy, and then the officer is given a short course dealing with the problems of the community in his or her own department.

Methods of Instruction

There are many different methods of instruction, but only the three used most often in police training courses are discussed below.

Imparting Information

Adequate and accurate information is essential for in-service training. Often, the only procedure necessary to "get over" a point is to merely *give* the facts—for example, special or general orders and changes in policies or procedures. In this type of instruction, the individual is allowed no opportunity to react to the instruction—he or she merely acquires the facts.

Course Work or Organized Instruction

Course work is absolutely essential for the officer who has no experience in a given field. This type of instruction presupposes that the instructor will use accepted teaching methods—particularly the four steps of instruction discussed in the following section. Such training is not always suitable for the officer who has had experience in the subject matter being covered by the course. For the experienced officer, the conference method of instruction is often more suitable, and it is by far one of the most practical and interesting methods.

The Conference Method

In this method of instruction, specific problems are set up for discussion and analysis by a relatively small group. Allen Gammage defines the conference as the "collecting, classifying, and evaluating of facts or opinions secured from a group of people."[5] A conference is actually a discussion directed toward a specific objective, which has the advantage of providing an interchange of ideas and experiences. Group enthusiasm may be engendered in such a conference, and group unity fostered and maintained. During a conference, the prejudices of a single point of view can be dissolved in the fluidity of a discussion probing varied points of view.[6]

It is the opinion of the authors that the conference method is so important to the police supervisor that all new supervisors should be given a specialized course in the techniques of leading a conference. Clearly, the conference leader can make the meeting either a success or a complete failure, depending upon what he or she does or fails to do. For a more complete exposition of the conference method of teaching, see Gammage's *Police Training in the United States*. The United States Air Force also has published an excellent training manual on conference leading which is available in some local Air Force base libraries. See Allen Bristow's *Police Film Guide* for sources on the film entitled, *All I Need Is a Conference,* an excellent film on conference leading.[7]

Four Steps of Instruction

In teaching any subject, it is essential that the police supervisor be thoroughly familiar with the steps of instruction that are essential to meaningful and effective teaching. The number of steps varies from one training program to another, but generally, it has been found that four steps are sufficient.

Introduction

The *Introduction* is used to explain to the student why it is important to learn the lesson, to get attention, to arouse curiosity and interest, and to develop an actual desire to learn—to make the student want to have the lesson taught. A number of methods may be used in the introduction, such as: (1) questions designed to catch the student's interest, bring out definite facts, and lead the student to the starting point of the lesson; (2) discussion; (3) citing cases; (4) informational demonstrations; and (5) illustrations.

Presentation

The second step is the *Presentation,* in which the material to be learned is presented to the student. Here the teacher gives the techniques, concepts, information, proper methods, etc., step-by-step. Methods used in the presentation may include: (1) demonstration, (2) explanation, (3) experimentation, (4) illustration, (5) questioning, and (6) lecturing.

Application

The third step is the *Application,* in which the student does the job or in some way uses the information, answers questions, or tries to solve problems. At this stage, the instructor assists the student if he or she gets stuck. This step is accomplished by: (1) letting the student do, under supervision, work which may include the use of charts to be filled in or large diagrams, individual forms, and reports, or possibly a bit of role-playing under the instructor's supervision; (2) letting students use the information by answering questions; (3) citing hypothetical cases and letting the students solve them; and (4) summarizing the information covered in the presentation and application.

Test

The fourth step is a *Test* in which the student either does the job or applies the knowledge without the help of the instructor. Properly applied, this test will help to determine the effectiveness of the instructor as well as the proficiency of the student. Testing is largely accomplished through: (1) application of what has been learned to the job, (2) the application of it to an artificial situation slightly different from what has been illustrated or cited, but in which the same concepts and principles or methods that have been taught can be utilized (this involves the use of actual or hypothetical cases that the student will solve without assistance), and (3) questions, which may be either written or oral. Testing must relate directly to the stated performance objectives.

Lesson Planning

An airplane built by skilled craftsmen with the finest materials available probably will not fly safely unless it is constructed according to a definite, carefully engineered plan. Likewise, a training course can be offered by a well-informed teacher and can contain excellent subject matter, but it is unlikely to be effective unless a definite and orderly plan is followed. Students learn better and more rapidly when instruction is properly planned. The simplest way to organize material is to make up a lesson plan so that the instructor always knows exactly where he or she is in the lesson and, more important, where the lesson is going.

One of the best procedures to follow in setting up and organizing a unit of instruction is to first write a set of performance objectives for the instruction. Performance objectives set forth the specific knowledge, information, skills, attitudes, etc. that the student is expected to achieve at the conclusion of the instruction, and they set up a measurable standard to assure that the objectives have been met. In the past, too much teaching has been done without first determining what it is that you want the student to know or be able to do at the conclusion of the instruction. This is, in a sense, "putting the cart before the horse," and it seems far more sensible to first determine what the end product is that you wish to produce and then design your instruction to accomplish that end result. An excellent and relatively short exposition on performance objectives may be found in the second edition of *Preparing Instructional Objectives* by Robert F. Mager, published by Fearon Publishers in 1975. Another good treatise on the same subject from the same publisher is *The Uses of Instructional Objectives: A Personal Perspective* by W. James Popham, published in 1973.

No matter how well a teacher knows a subject, time must be spent arranging the subject matter in the proper manner to effectively communicate it. With the objective of the lesson in mind, the teacher selects only the essential points about the subject and outlines them for the most effective presentation. With the lesson carefully outlined, the instructor is in a position to organize details so that no time is wasted in the class session. A suggested format for lesson plans is shown in Figure 9. Figure 10 is an illustration of a completed lesson plan.

When an assignment is to be given, a thorough instructor writes exact directions into the lesson plan. Assignments should be so clear that they cannot be misunderstood. One way to avoid misunderstanding is to issue an *assignment sheet* supplying all of the necessary reference information, including the page numbers for recommended reading. There should be no excuse for students not being able to locate the required sources. A properly developed lesson plan enables the instructor to provide students with a sound basis for practically applying what they have learned. The details may vary, but the main divisions are usually the same. The lesson plan is a systematic, businesslike way to organize the subject matter and is well worth the time it takes to prepare. It can make or break the teaching job.[8]

Some Principles of Teaching

The police instructor must be aware of the principles involved in teaching effectively, from the standpoint of teaching techniques and of the psychology of learning. Many capable police officers with little or no experience in teaching are suddenly called upon to act as instructors upon promotion to a supervisory rank. Too often, they do not have adequate time to prepare themselves for their new responsibilities. Most of them do not have sufficient time to study any of the many excellent books that are available on the principles of teaching. For this reason, the following brief guidelines may help the police instructor to teach effectively.

Characteristics of a Good Teacher

Directing people properly requires special personal attributes; more teachers fail to teach effectively because of personality problems than because of a lack of job knowledge and skills. Of course, teaching requires more than the specific skills that are necessary to conduct an investigation or drive a patrol car at high speeds. A teacher does not direct a student's thoughts merely by telling the student what to think; rather, the teacher guides the student by stimulating interest in a subject and by encouraging a desire to learn. If a teacher misses the target on just one major point, the student may be totally misdirected.

ANYTOWN POLICE DEPARTMENT RECRUIT SCHOOL LESSON PLAN

COURSE: Title of course.

LESSON: Title of specific lesson — short, complete, and descriptive of the lesson contents.

PERFORMANCE OBJECTIVES: List the knowledge, appreciations, skills, and abilities the student will gain from this lesson.

MATERIALS NEEDED: List all of the materials that will be given to the students and the supplies, equipment, and visual aids you will need to teach the lesson.

ASSIGNMENT: Outline any student assignments and, if you expect outside study, hand out assignment sheets.

REFERENCES: List all of the specific references from which you developed this lesson.

INTRODUCTION: Write out what you plan to do to prepare the students for the information you are going to give them and how you will arouse attention, curiosity, interest, and create a desire to learn. Tell them WHY this lesson is important to them.

PRESENTATION: List the teaching points in a logical step-by-step sequence in outline form, including all key points, examples, and stories to illustrate the points. In this step, tell how, show how, do the job, and present the information.

APPLICATION: List what you will do and the techniques you will use so the students may apply the knowledge you have given them in the presentation. People learn by doing, and in this step the students do the job or apply their new knowledge under the teacher's direction. Ask questions and summarize the points covered in the presentation.

TEST: List the testing techniques you will use — performance, oral, or written, depending upon the type of lesson. Specify exactly what students will do, and include a copy of any written tests. Testing must be directly related to the stated performance objectives.

FIGURE 9. A Suggested Format for Lesson Plans

ANYTOWN POLICE DEPARTMENT RECRUIT SCHOOL LESSON PLAN

COURSE:	Patrol Procedures.
LESSON:	How To Make a Preliminary Search of a Suspect.
PERFORMANCE OBJECTIVES:	1. Recruits will be able to recognize the dangers involved in stopping and searching suspects. 2. Recruits will be able to demonstrate proper and safe techniques for stopping and searching suspects. 3. Recruits will be able to demonstrate the proper use of gun and handcuffs.
MATERIALS NEEDED:	1. Slide projector and slides, projection screen. 2. UNLOADED service revolver and holster. 3. Handcuffs. 4. Display of weapons taken from suspects.
ASSIGNMENT:	None.
REFERENCES:	International Association of Chiefs of Police TRAINING KEY
INTRODUCTION:	1. Explain need for search and that it may cost life or cause serious injury. 2. Cite recent cases in which officers were injured or lost prisoners by not searching properly. 3. Point out that loss of prisoner will bring censure to officer and department. 4. Ask class if they would like to know a "good method" of searching. Tell them you will show how to search a prisoner. 5. Show slides of officer on ground with armed suspect standing over the officer.
PRESENTATION:	1. Explain and demonstrate most common places where weapons are concealed — use student as subject. 2. Apprehend suspect from rear (one method). Advise to use caution — not to get careless. 3. Draw gun — stop suspect — say "hands up and don't move" — identify yourself as a police officer. 4. Brace suspect against a wall or vehicle. Show slides of approach and proper positioning of suspect.

FIGURE 10. Sample Lesson Plan

5. Search suspect's hands first — officer searching with left hand, gun in right hand.
6. Show slide of search position and explain.
7. Show slide of position of officer's feet and explain importance.
8. Explain how to complete the search — use slides — advise extreme caution.
9. Show how to secure and place handcuffs before suspect is allowed to straighten up from wall — show slides of position of suspect's hands.
10. Suspect can now be straightened up, ready for transportation — officer's gun holstered.
11. Demonstrate the entire procedure using one of the students as a suspect.

APPLICATION:
1. Demonstration by class members, with aid of instructor. Two at a time, one as officer and one as suspect, then have them reverse roles.
2. Bring in a previously prepared "suspect" who is carrying at least ten different weapons and have suspect searched by a student. Then show what weapons were missed and why.
3. Questions and class discussion.

TEST: Set up a slightly different situation of meeting the suspect and making apprehension from the front — have students demonstrate with no aid from the instructor.

FIGURE 10, (cont'd.) Sample Lesson Plan

Even Disposition   A good teacher must possess an even disposition and cannot afford to allow the peculiarities of individuals to be disturbing. In order to secure the best efforts from students, a teacher must always be fair, firm, and friendly. No matter how much difficulty a student has in understanding a problem, the good teacher is always patient.

Bearing and Appearance   A teacher's bearing and appearance play a definite role in his or her effectiveness. Bad posture gives the impression that the teacher lacks self-confidence, and students pay less attention to unassured instructors. Gestures can be used to emphasize a point dramatically and to keep the attention of the class from wandering. A teacher should have vitality, but should remember that small, incomplete gestures distract rather than command attention. Therefore, gestures should be used sparingly, but when they are used they should be meaningful. If a teacher dresses in a slovenly manner and exhibits poor conduct, attention is directed to the person rather than to the subject matter. The teacher should always be well-groomed, with clothing freshly cleaned and pressed, shoes polished, and brass shiny. Civilian clothes should be conservative in design and color.

Good Habits   The teacher must set the example for the class to follow. He or she must start and finish the class promptly. A good teacher is systematic and thorough, organizing the lesson carefully and moving methodically from point to point.

Diction   Because the supervisor gives most of the instruction verbally, he or she must pay a good deal of attention to speech. If the teaching supervisor fails to speak clearly, using good enunciation and pronunciation, the attention of the students is bound to wander. If he or she speaks too rapidly, students will not understand what was said. If the teacher speaks in a monotone, it is not unreasonable to expect the students to fall asleep, particularly if the class follows a meal or if some of the students have just finished their shift. If the supervisor speaks to the blackboard, it is not reasonable to expect intelligent "feedback." The speed, pitch, and tone of voice should be controlled carefully. If the instructor does not speak with authority, the students will probably dismiss much of what is said as unimportant.

Timing is another important factor in verbal instruction—the pause, the studied saying of significant words and phrases, and the casual aside are useful devices. Poor timing leaves the student with a vague impression, while proper timing gives him or her a clear picture.

Enthusiasm   The police supervisor who must teach subordinates has a tremendous responsibility. He or she must work with the most valuable and, at the same time, the most plastic material in existence—the human personality. For this reason, the supervisor must approach this phase of the job with enthusiasm and a strong sense of responsibility. A good instructor is always alert for new and interesting ways to attract and hold students' attention, and receives a strong sense of personal satisfaction from arousing students' curiosity and making them eager to learn. He or she constantly strives to learn more about the subject than is required in order to teach. The more the instructor knows, the more the class will be stimulated.

A primary goal of all supervisors is to help subordinates to understand and to learn. In order to accomplish this, the supervisor must first inspire subordinates to want to learn. He or she can do this best by making the subject interesting.

Discipline There is one more important requirement for effective teaching, and that is discipline. If the teacher is fair, firm, and friendly, discipline is not difficult. The instructor must insist on good conduct and punish any student—if only with a mild reprimand—for improper behavior. A word of warning, however, is that, with very few exceptions, reprimands should be given in private, *not* before the class. When students feel that they are receiving a square deal, they seldom cause trouble in class, regardless of how strict the discipline is. In fact, students seem to prefer firmness as long as it is accompanied by fairness. Friendliness results in pleasant working conditions and greater accomplishment. However, friendliness should not be confused with familiarity; the latter weakens authority.

Needs of the Student

From the viewpoint of the student, good teaching techniques are essential. The student wants a teacher to inspire him or her to learn. Whether the student is going to become a traffic officer, a helicopter pilot, a detective, or all three, he or she needs to know what to achieve and how to do it. For some students, the desire to learn is the reason they are in a class. Too often, unfortunately, the student in a police training course is only there because he or she has to be, not because of a specific goal. However, the student without a goal often unconsciously wants one and will respond to a good instructor. All students like an instructor who can present the subject in an interesting manner and make routine subject matter come to life.

Context Students generally want to know how a subject fits into the overall scheme of things, and they are more eager to learn if a subject's significance is clear to them. The student needs to learn one thing at a time, beginning with simple concepts and progressing gradually to the more complex facts of the problem. A recruit does not begin by investigating complicated homicide cases, but rather by progressing to this level gradually, acquiring the needed knowledge and skills as he or she gains experience in the field. A good instructor gives students an overall picture of the subject at the beginning of the course.

Continuity   It is extremely helpful to the student if an instructor relates each new concept to the material that has been learned. For example, unless the student already knows the elements of the crime of robbery, there is no point in talking about case law and court decisions pertaining to robbery. The student wants learning to be a process of fitting each new fact or concept into the structure of knowledge he or she has already built.

Reinforcement   Few people can remember a fact after hearing or reading it only once. Information must be reinforced in some manner to make a more permanent impression and to facilitate memory. A new police officer can study a book on interrogation, memorize most of the material, and through study and review, achieve a vague understanding of the subject. However, only after participating in actual interrogations will the young officer fully understand the principles involved and develop the self-confidence that is necessary to do an effective job. In other words, the student needs *review* and *practice* to reinforce learning.[9]

Speeding Up the Learning Process

Part of an instructor's job is to give the student as much information as possible in the shortest time possible. No police agency can afford an inefficient use of time and personnel. A few of the most effective methods to shorten the learning process are covered in the following discussion.[10]

Use Experts for Instructors   The instructor must know the subject thoroughly. It is not possible to be an effective teacher if you stay only one jump ahead of the class. The good instructor is able to answer almost any question about the subject immediately—not at the next class meeting after looking up the answer. The use of a properly prepared lesson plan keeps the lesson headed in the right direction, without unplanned diversions from the topic, and keeps the students progressing in easy steps and at the right pace. Lesson planning was discussed in more detail earlier in this chapter.

Maintain Student Interest   Whether in the classroom or in field training, it is imperative that the instructor arouse and maintain the students' interest. When the teacher steps before a class, some of the students will be thinking about things far removed from the subject matter of the lesson—one student may be thinking about a new car, while another is still munching on the thick, juicy steak sandwich he or she had for lunch. If the instructor fails to arouse interest in the lesson immediately, the car will completely capture the first student's interest, while the second student will progress to dessert. Interest can be aroused by the instructor's enthusiasm if it is sincere.

However, false enthusiasm is quickly felt by students, and it will leave them cold.

Relating personal experiences is another technique for arousing interest in the subject matter. However, the teacher must not allow the class to become a forum for his or her entire personal history; personal experiences should be brief and to the point—sufficient to arouse interest but not unrelated to the lesson. The instructor can also arouse interest by connecting a new lesson to the past experiences or future needs of the students.

Keep It Simple   Another method of effective teaching is to keep the instruction as simple as possible. The teacher should use simple words—unless complex ones are really necessary—and should explain complicated or technical terms when they are used. The supervisor who persists in using—and, in too many cases, misusing—long words, soon loses the respect of subordinates and becomes the butt of many jokes. Another way to keep things simple is to explain complicated matters in terms of everyday things that the students understand. An example is to explain the polygraph's recording of physiological changes in terms of the sensations a person feels when placed under stress—the mouth becomes dry because of inhibition of the salivary glands, the heart pounds, breathing is faster, etc.

The instructor should present one idea at a time in logical steps when introducing new material. It is easier for a student to understand a new concept if the teacher is able to show the whole picture and explain the relation of one idea to another.

Keep Them Thinking   One of the most important aspects of teaching is to keep the students thinking, and a good method of accomplishing this is to ask questions. An answer requires thought, and if students do not know when the instructor will call on them, they are forced to think about each question that is raised. This method is most effective if the student's name is called *after* the question is raised. A discussion also helps students to think; when they want to argue a point, it's a good sign that they are thinking about it. However, the discussion must pertain to the subject and should not be allowed to wander off the track. Good discipline will also help students to think. It is human nature to work harder when we know it is expected of us. Perhaps the most important way to keep students thinking is to have them *do* what they have been taught. This method is particularly important when *skills* of varying types are being taught. It simply is not possible to teach how to roll a set of fingerprints without having the student do it. There is no learning without either physical or mental activity—preferably both. People remember best what they *do,* not what they *see* or *hear.* Also, this method helps the instructor evaluate the students' performance to determine if they have learned what they were supposed to learn.

Audio-Visual Teaching Aids

The instructor should use all reasonable teaching methods that appeal to the eyes and ears of the students. Audio-visual teaching aids are excellent for arousing and maintaining interest, as well as for clarification of difficult material. A number of excellent police training films, slides, and other aids are available from various sources, such as state Peace Officers' Commissions, the International Asociation of Chiefs of Police, state film libraries, local police agencies, and college and university audio-visual libraries. Any supervisor who is assigned teaching responsibilities of any kind should be familiar with local sources of teaching aids.

Some of the most effective aids are those that are created by the instructor to demonstrate or illustrate the various points of the subject matter. For example, the laws of various states that pertain to the control of deadly weapons often use terms such as dirk, dagger, stiletto, switchblade, gravity knife, etc. A display board with examples of each weapon will do more to clarify definitions than hours of explanation by the instructor. The supervisor should be familiar with whatever audio-visual equipment is available in the department and learn how to operate it. He or she will need to know the differences between an opaque projector and an overhead projector and how and when to use each of them, and must be able to operate a motion picture sound projector, filmstrip and slide projector, tape recorder, etc. One of the newer aids that is readily adaptable for use in teaching, particularly in the police agency, is the videotape recorder that is used in conjunction with a closed circuit television monitor. A number of police agencies are currently using this equipment for training and other purposes. Manufacturing costs of the videotape recorder have now been reduced to the point where it has become practical even for those departments with relatively limited budgets. There are many advantages to this type of equipment, particularly its instant playback capability. For example, a student could be asked to demonstrate an interrogation technique and, immediately following the demonstration, could see and hear a playback which would pinpoint the good and bad points of the demonstration.

An instructor should *show* students what he or she is talking about by using pictures, models, and actual demonstrations. What students see and participate in its remembered much better than what they hear.

Two devices used by many police instructors are the *Field Instructor's Schedule*[11] and the *Field Training Guide*.[12] They serve as a check-list for field instructors by providing a breakdown of the various police field operations that need to be taught to new police officers. The *Field Training Guide* is reproduced in Appendix I.

Another useful device for the supervisor is some form of a training progress chart which shows at a glance the status of officers in the current training program. See Figure 11 for a sample of the chart. As a new officer

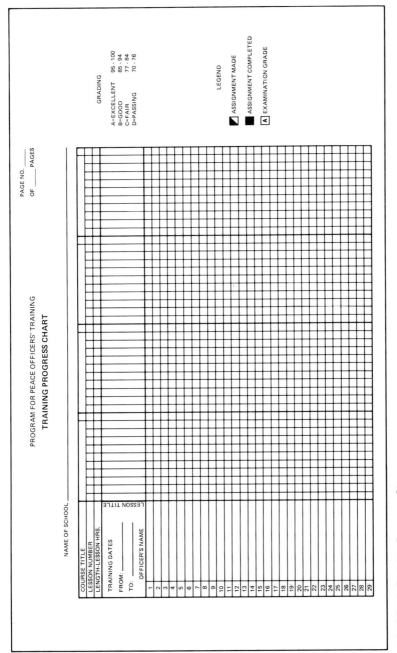

FIGURE 11. Training Progress Chart

learns how to perform each particular task satisfactorily, the supervisor checks that job off the chart. The training progress chart was developed by the Wichita, Kansas, Police Department many years ago and has since been modified and used by many other police agencies.

In summary, it can be said that the training and education function of the police supervisor is one of the most important aspects of supervisory work. In this capacity, it is the supervisor's responsibility to train new officers, to provide advanced training for experienced personnel, and to serve as a counselor on training and education for all who need assistance. In this teaching role, the supervisor must be familiar with what is educationally available to subordinates, both in and out of the department, and must be a good teacher, able to inspire officers to want to better themselves professionally. The supervisor is the one person to whom officers must be able to look for guidance as well as leadership.

DISCUSSION QUESTIONS

1. Why is the supervisory training and education function growing in importance?
2. What are the most important features of Peace Officers' Standards and Training legislation?
3. What are the differences between education and training?
4. Who is responsible for determining the training needs within a department?
5. What is the supervisor's responsibility for instruction?
6. What methods can be used to determine training needs?
7. What are the various phases of training in the police service and what does each involve?
8. What are the four principal characteristics that are essential to efficient training programs? Explain each of them.
9. Why is the proper orientation of recruits so vital?
10. How is recruit training made interesting for the new officer?
11. What is the importance of field instruction, and whose responsibility is it? Why?
12. What are the important aspects of an induction interview? Why are they important?
13. What is the purpose of advanced officer training?
14. How can the supervisor stimulate interest in advanced training and education?
15. What incentives can a department employ to encourage officers to advance their education?
16. What sources outside the department can be used for training of personnel?
17. Describe the three most commonly used methods of instruction. Describe some of the chief advantages and disadvantages of each one.

18. Describe in detail each of the four steps of instruction.
19. What is the importance of lesson planning?
20. How should assignments be made?
21. What are the attributes of a good teacher? Explain each one in detail.
22. What are some of the most important causes of teacher failures? Why?
23. How does a good teacher maintain discipline in class?
24. How does an instructor arouse interest and motivate students?
25. What techniques are used to reinforce the learning process?
26. How does a teacher encourage students to think?
27. Discuss the various types of teaching aids. How can each of them be utilized?
28. What is the use for the *Field Training Guide?*
29. What are the advantages of a training progress chart?
30. Discuss the supervisor's role as a training counselor.

NOTES

1. William B. Melnicoe and John P. Peper, *Supervisory Personnel Development* (Sacramento, Calif.: California State Department of Education, 1965), p. 120.
2. Ibid., p. 121.
3. Melnicoe and Peper, op. cit., p. 122.
4. Melnicoe and Peper, op. cit., p. 125.
5. Allen Z. Gammage, *Police Training in the United States* (Springfield, Ill.: Charles C. Thomas, 1963), p. 238.
6. Ibid., pp. 238-239.
7. Allen P. Bristow, *Police Film Guide,* 1968 Issue, available from Police Research Association, P.O. Box 1103, Walteria, California 90505, for $5.00.
8. Derald D. Hunt, *Teacher Training Supplemental Material No. 27* (Peace Officers' Training Program, California State Department of Education, 1960), pp. 1-4.
9. *Teacher Training Supplemental Material No. 156* (Peace Officers' Training Program, California State Department of Education), pp. 1-4.
10. *Teacher Training Supplemental Material No. 154* (Peace Officers' Training Program, California State Department of Education), pp. 1-5.
11. O. W. Wilson, *Police Planning* (Springfield, Ill.: Charles C. Thomas, 1952), pp. 314-321.
12. *Field Training Guide* used by the Berkeley, California, Police Department is reproduced with permission in the Appendix.

Appendix A

**Any City, U.S.A.
Police Department**

Example of a General Order:
Complaints of Police Personnel Misconduct

Purpose The purpose of this order is to insure the integrity of the police department by establishing procedures that will assure the prompt and thorough investigation of alleged or suspected misconduct. Such procedures will:
   (1) Clear the innocent.
   (2) Establish the guilt of wrongdoers.
   (3) Facilitate prompt and just disciplinary action.
   (4) Uncover defective procedures or material.

Procedures
   (1) All alleged or suspected violations of law, ordinances or department rules, regulations or orders, must be investigated. These include:
      A. Those violations reported to supervising or commanding officers by:
         1. Members of the department, either orally or in writing.
         2. Citizens (including prisoners), in person, by telephone or correspondence, either signed or anonymous.
      B. Those violations observed or suspected by supervising or commanding officers.
   (2) The term "misconduct," as used in this order, shall be defined as: "A violation of any law, written department policy, rule or order, or immoral conduct or behavior unbecoming an officer, committed whether on or off duty."

GENERAL ORDER: COMPLAINTS OF POLICE MISCONDUCT   257

(3) Upon receipt of information of alleged misconduct on the part of any member of this department, the employee receiving such information shall refer the matter to his or her supervisor or commanding officer for investigation.

(4) A commanding officer receiving any complaint of misconduct shall summarize the information of the complaint and forward a written copy of the summary to the Chief of Police and to the Division Commander of the person or persons against whom the allegation is made. The personnel complaint report form may be used to serve as a first report for any incident reported. (See sample.) Any reference to

```
ANY CITY POLICE DEPARTMENT
PERSONNEL COMPLAINT REPORT

┌─────────────────────────┬──────────────────────┬──────────────────────────────┐
│ Date & Time of Occurrence│ Location of Occurrence│ Date & Time Reported to Police│
├─────────────────────────┼──────────────────────┼──────────────┬─────┬─────────┤
│ Name of Complainant     │ Residence Address    │ Residence Phone│ Age │ Sex    │
├─────────────────────────┼──────────────────────┼──────────────┴─────┴─────────┤
│ Business Name           │ Business Address     │ Business Phone              │
├─────────────────────────┼──────────────────────┼──────────────┬─────┬─────────┤
│ Witness (Name)          │ Residence Address    │ Residence Phone│ Age │ Sex    │
│ 1.                      │                      │              │     │         │
│ 2.                      │                      │              │     │         │
├─────────────────────────┼──────────────────────┼──────────────┼─────┴─────────┤
│ Name of Officer(s)      │ Description of Officer│ Badge No.    │ Uniformed     │
│ (If Known)              │                      │              │ Officer (Check)│
│ 1.                      │                      │              │               │
│ 2.                      │                      │              │               │
└─────────────────────────┴──────────────────────┴──────────────┴───────────────┘
Describe details of occurrence (Use reverse side if additional space is needed).
_____
_____
_____
_____
_____
_____
_____

┌──────────────────────────────────────────┬────────────────────────────────────┐
│ I hereby declare under penalty of Perjury│ Witness to Affidavit _____ │
│ that the foregoing is true and correct.  │                                    │
│ By: _____         │ _____   _____  │
│       Complainant's Signature            │ Residence Address Residence Phone  │
│ _____                 │ _____   _____  │
│     Date and Time Signed                 │ Date & Time Signed Age of Witness  │
├──────────────────────────────────────────┼────────────────────────────────────┤
│                                          │ FOR DEPARTMENTAL USE ONLY          │
│ AUTHORITY: Section 446 Civil Code of     │ Report received by Officer_____ │
│            Procedure                     │                                    │
│ PENALTY : Section 23-8 Any City Municipal│ Date & Time Received _____ │
│            Code.                         │                                    │
└──────────────────────────────────────────┴────────────────────────────────────┘
```

the Chief of Police shall include his or her authorized representative, the Assistant Chief, and in the Assistant Chief's absence, the on-call Division Commander to whom notification must be made.

(5) If, in the opinion of the officer in charge at the time the complaint is received, the incident is of sufficient gravity to demand immediate action, he or she shall notify the commanding officer, regardless of the hour. In addition, he or she shall take any immediate action necessary to preserve the integrity of the department until the arrival of the commanding officer.

(6) Upon review of the summarized report of an allegation of misconduct, the Chief of Police will assign the matter to a Staff Officer, who will be responsible for the investigation. Investigation responsibility will normally be assigned to the employee's Division Commander.

(7) The officer assigned to investigate an alleged act of misconduct on the part of a member of this department shall conduct a thorough and accurate investigation. Such investigation shall include formal statements from all parties concerned, when necessary and pertinent, and the gathering and preservation of any physical evidence on the matter.

(8) Complaints by citizens against members of this department shall be processed in the following manner:
   A. During normal business hours such complaints shall be referred to the employee's Division Commander or the Commander's representative.
   B. At times other than normal business hours, such complaints shall be referred to the officer in charge, who shall take the necessary and appropriate action.

(9) When the investigation is completed, the final report will conclude with the classification of the investigation into one of the following categories:
   A. *Unfounded*
      Allegation is false or not factual.
   B. *Exonerated*
      Incident occurred, but was lawful and proper.
   C. *Not Sustained*
      Insufficient evidence either to prove or disprove.

(10) When the act complained of is a crime and the evidence is such that had the action been by a private person it would have resulted in arrest, the investigating officer will explain the circumstances to the commanding officer or Chief of Police and request a decision as to whether:
   A. The accused person should be arrested forthwith; or
   B. A warrant for the accused's arrest should be first obtained; or
   C. Criminal action should be delayed pending further investigation; or

GENERAL ORDER: COMPLAINTS OF POLICE MISCONDUCT   259

   D. The accused shall be taken into "protective custody" pending a decision by the Chief of Police.
(11) When a member of this department has been arrested by this department or any other jurisdiction, the Chief of Police shall be notified immediately.
   A. The Chief of Police shall instruct the Commanding Officer to proceed to the place of confinement and advise the employee that he or she is suspended immediately, pending further investigation.
   B. An immediate investigation shall be conducted by the assigned Staff Officer and the results communicated to the Chief of Police.
   C. If the arrest is unwarranted, the officer will be reinstated without loss of pay or benefits.
   D. If the arrest is warranted, appropriate action shall be instigated immediately.
(12) The Chief of Police shall be constantly apprised of the status of each personnel investigation and upon completion of such an investigation shall review each case and the recommendations of the investigating officer and take appropriate action.
   A. The complainant and the officer concerned shall be notified of the results of the investigation and the disposition of the case in writing.
   B. Copies of all documents concerning the investigation shall be preserved in the employee's personnel file.
   C. Contents of personnel jackets containing such investigations shall not be released to persons outside the department without the express approval of the Chief of Police, and such release must be in strict compliance with existing statutes.

Appendix B

Oakland Police Department Complaint Form

GLUE FLAP DOWN TO MAIL

WRITE YOUR NAME, ADDRESS AND PHONE NUMBER

WRITE THE DATE THIS FORM IS FILLED IN   WRITE THE DAY & DATE OF INCIDENT OR ACTION   TIME OF INCIDENT

WHERE DID THE INCIDENT OR ACTION TAKE PLACE?

WRITE THE NAMES OF ANY WITNESSES, THEIR ADDRESSES AND TELEPHONE NUMBERS

IF A PERSON WAS ARRESTED, WRITE HIS OR HER NAME, ADDRESS AND PHONE NUMBER IF KNOWN

IF A POLICE OFFICER WAS INVOLVED, WRITE HIS OR HER NAME, BADGE NUMBER AND CAR NUMBER, IF YOU HAVE THIS INFORMATION

WRITE THE NATURE OF OPINION OR COMPLAINT

OAKLAND POLICE DEPARTMENT
Police Administration Building, 455 7th St
Oakland, California 94607

To the People of the City of Oakland:

I wish to assure you that your Police Department welcomes constructive criticism of Department procedures or valid complaints against police officers. Each criticism and complaint received will be investigated thoroughly and appropriate corrective action taken when warranted by the facts obtained. You will be informed of the results of the investigation when it is completed.

If you wish to make a complaint, you may come to my office on the 8th floor of the Police Administration Building, at 7th and Broadway, or phone 273-3365. You will be received by my Administrative Aide or by the Sergeant in charge of the Internal Affairs Section, whose job it is to investigate complaints. You will be treated courteously and thorough consideration will be given to the problem you present.

If you do not wish to come to the Police Department or telephone, you may register your complaint in writing to me. Just complete the form, seal the glued flap and mail this pamphlet. I am the only one who will open it. Please write as much information as you can; it will be helpful if you will give your name and address so that we may contact you for further information if needed. Any information you give will be kept confidential if you so request.

Please feel free to express yourself on any problem which you feel should be directed to my attention.

Sincerely,

C. R. Gain
Chief of Police

Appendix C

Personnel Evaluation Form

NAME								
Last	First	Middle Initial	P No.	Rank	Date of Rank	D.O.B.	From	To

TYPE OF REPORT:	UNSATISFACTORY	NEEDS IMPROVEMENT	COMPETENT	ABOVE AVERAGE	OUTSTANDING	BASIS OF EVALUATION:	Date From	Date To
☐ Cadet						☐ Direct supervision		
☐ Probationary Period # _____						☐ Close duty contact		
☐ Annual						☐ Frequent observation		
☐ Transfer						☐ Infrequent observation		
☐ Interim						☐ Reports and records		
							FILL IN ABOVE IF NOT FULL PERIOD	

RATE EACH APPLICABLE SECTION						Sec. No.	COMMENT — Required on ratings of "Outstanding" "Needs Improvement" or "Unsatisfactory".
1. INITIATIVE							
A. Consider ability to initiate police action.							
B. Does he or she pursue investigative leads to the utmost?							
C. In grading a supervisor, consider willingness to make decisions, laxity or willingness in discussing pertinent matters with those supervised.							
D. Does he or she put things off?							
E. Does he or she perform with a minimum of supervision?							
2. COOPERATION AND ATTITUDE							
A. Ability and willingness to work in harmony with and for others.							
B. Acceptance of authority and supervision.							
C. Use of tact with employees and the public.							
D. In grading a supervisor, consider willingness and practice in receiving and passing on orders, details, follow-up procedures.							
3. JUDGMENT							
A. Ability and practice in doing and saying the right things.							
B. Do his or her actions tend to bring credit to the department?							
C. The ability to think of many things at once, in their interdependence, their relative importance and their consequences.							
D. Consistent exercise of good, independent judgment.							
E. Knowledge of strengths and weaknesses of subordinates and, as a result, obtains the greatest efficiency through them.							

NAME								Sec. No.	COMMENTS
Last		First	UNSATISFACTORY	NEEDS IMPROVEMENT	COMPETENT	ABOVE AVERAGE	OUTSTANDING		

4. RELIABILITY

   A. Consider truthfulness, loyalty and dependability.
   B. Is he or she consistent?
   C. Has he or she a strong interest in the work?
   D. Thoroughness.

5. TIME FACTORS

   A. Punctuality.
   B. Makes good use of time.
   C. Works to full capacity.
   D. Written reports submitted on time.
   E. Consider willingness to use own time to benefit the department.

6. PERSONAL APPEARANCE

   A. Neat?
   B. Clean of person, uniform, other working clothing?
   C. Does he or she always meet or exceed the standards of neatness and grooming?

7. SAFETY CONSCIOUSNESS

   A. Consider needlessly endangering him or herself or others.
   B. Regard for good practices in car stops, suspect contacts, prisoner handling.
   C. Safe driving habits.
   D. Care and effective use of department property.
   E. Total consideration for safety — benefit to the rated employee, fellow employees, citizens, welfare of the city and city equipment.

8. QUALITY OF WORK

   A. Consider knowledge and application of laws, ordinances, general orders and standards of operating procedures.
   B. Investigation and reporting techniques.
   C. Clean, well done job?
   D. Ingenuity, resourcefulness.
   E. Compliance with instructions.
   F. Consideration of departmental objectives.

Personnel Evaluation Form (Cont.)

PERSONNEL EVALUATION FORM

263

NAME							Sec. No.	COMMENTS
Last		First	UNSATISFACTORY	NEEDS IMPROVEMENT	COMPLETE	ABOVE AVERAGE	OUTSTANDING		

9. QUANTITY OF WORK

- A. Aggressiveness?
- B. Amount of work done.
- C. Consideration for departmental objectives.

10. SUPERVISORY ABILITY

- A. Accepts responsibility?
- B. Delegates work?
- C. Maintenance of high morale?
- D. Training and instructing?
- E. Fairness and impartiality?
- F. Approachability?

11. POTENTIAL FOR INCREASED RESPONSIBILITY

- A. Training.
- B. Experience.
- C. Demonstrated potential.

12. PROBATIONARY RECOMMENDATIONS

- ☐ Permanent Status
- ☐ Further Observation
- ☐ Termination

13. OVERALL RATING

- Top 10% _____
- Upper 25% _____
- Middle 50% _____
- Lower 25% _____
- Bottom 10% _____

	Signatures of Raters	Rank	Date
This report has been discussed with me			
	Signature of Division Commander	Rank	Date
	This report discussed with employee	Rank	Date
Employee's Signature Date	By:		

Personnel Evaluation Form (Cont.)

Appendix D

**Any City, U.S.A.
Police Department**

Example of a General Order: Permanent Employment
Status—Achievement

Purpose The primary objective of the employee selection process is to provide the best possible candidate for the available job. In addition to a rigorous screening program before employment, the new employee must successfully complete a period of probationary employment. To insure the critical evaluation of the employee during the probationary period, the following criteria are established.

Policy Once hired, the employee is on a probationary employment status for one year. To gain permanent status a police employee must satisfactorily meet the established standards of appearance, compliance, and performance.

Standards The following comprehensive program of evaluation shall be applied in review of recruit police officers to determine their suitability for recommendation as a permanent status employee.

I. GENERAL CONDITIONS
   A. These general criteria will be used as guidelines in evaluating the recruit during the probation period.
      1. Presents a neat appearance.
      2. Exhibits acceptable character traits and moral attitudes.
      3. Demonstrates:
         a. Emotional stability.

GENERAL ORDER: PERMANENT EMPLOYMENT STATUS 265

      b. Mental maturity and intelligence.
      c. Sound judgment.
      d. Personal integrity.
      e. Courage.
    4. Ability to work free from conflict with fellow employees and the public.
    5. Ability to comprehend and apply extensive training and detailed instruction.
    6. Ability to conform to organizational objectives.
  B. To effectively record the impressions, opinions, and attitudes of the recruit's supervisors during the probationary period, an Evaluation Report shall be made monthly on each recruit.

II. TRAINING

The initial months of employment will encompass a lengthy training period for the recruit officer, which will be administered in three phases:
  1. Indoctrination Training
    Indoctrination Training will commence immediately upon employment, and will continue until the recruit is assigned to a Training Academy.
  2. "Basic" Academy Training
    "Basic" Academy Training will commence upon completion of the Indoctrination Training Program, and will continue for the duration of the academy program.
  3. Practical Field Training
    Practical Field Training will be administered in two phases. Phase One will comprise a large part of the final phase of Indoctrination Training, during which the recruit is assigned to the Uniform Division. During this period he or she will be assigned to field duty as the *passive* partner in a patrol unit, under the guidance of a training officer. Phase Two will comprise a period of time (depending upon the individual achievement) after graduation from the Basic Academy, during which the recruit will be assigned to field duty as the *active* partner in a patrol unit, under the guidance of a training officer.

III. INDOCTRINATION TRAINING

  A. This phase of the program is designed to provide a comprehensive introduction to the overall department operation. This program will last for a three-month period, and will involve the following assignments for the recruit.
    1. Service Division (3 weeks)
      a. First Week—Assigned training officer provides familiarization with personnel, facilities, equipment, and basic job knowledge. In addition, indoctrination to departmental regulations, methods of reporting, and procedures will be covered.

b. Second Week—Under direct supervision, the recruit will work in the Records and Communications sections with assigned personnel to gain a general familiarization with office, records, and communications procedures.
c. Third Week—Under direct supervision of the duty Dispatcher, the recruit will be assigned to operate the radio-desk and telephone switchboard.
d. During this three-week period, the recruit officer shall spend at least two days at the Police Range.
   (1) The Range Officer will provide weapons instruction, with emphasis on safety and on policy controlling weapons use.
   (2) Recruits shall receive both practical demonstrations and personal experience in primary weapons (pistol, chemical agents, and shotgun) use and handling.
   (3) Recruits are not allowed to carry weapons until completion of the program of weapons familiarization and training, and until they have demonstrated a satisfactory comprehension of weapons safety with a minimal proficiency score on the Practical Pistol Course.
e. During this training period, an evaluative summary should be made on the recruit by the training officer and reviewed weekly with the Uniform Division Commander. Brief tests of the recruit's progress through the Indoctrination Training shall be administered weekly. Passing scores are an essential requirement for promotion to the next training phase.
f. At the end of this three-week period, the recruit's Evaluation Reports, test results, range scores, and general progress shall be reviewed by the training officer and the Uniform Division Commander to determine if areas of deficiency exist.
   (1) Deficiencies will be brought to the attention of the recruit, and methods of improvement recommended.
   (2) A determination will be made recommending continuation in the program or the termination of the employee.

2. Investigation Division (2 weeks)
   a. Fourth Week—The training officer will provide an orientation to the Detective Bureau and Juvenile Bureau functions, and a familiarization with basic division operations. Instruction will be provided on Court procedure, and other agency relationships.
   b. Fifth Week—The training officer will provide indoctrination to the elements of evidence, protection, collection, and preservation. Familiarization will be furthered by the recruit reviewing crime reports as a case study method of instruction.
   c. Time will be provided from this schedule for at least one day of training at the Police Range.

d. The same evaluation and review process will be observed during this second period of training as described in Section III, 1, e and f.
3. Uniform Division (7 weeks)
   a. The recruit will be assigned to duty for specific periods of time on each duty shift. He or she will be assigned as the passive partner in the patrol unit, under the direct supervision of a training officer. The training officer will provide instruction in the day-to-day routine of Uniform Division duties.
   b. The recruit will be evaluated and tested each two weeks on his or her progress in assignments, rather than on a weekly basis. Other than this exception, the same evaluation and review process, as described in Section III, 1, e and f, will be observed.
   c. During this training period the recruit will receive instructions, although basic, in the majority of departmental functions. The recruit's ability to perform the duties of a police officer can amply be judged during this period, and will be recorded by the Uniform Division Commander. If the recruit lacks the desirable abilities, or for some mental or physical reason is unable to perform ably, the Uniform Division Commander shall record such facts.
   d. A report prepared by the Uniform Division Commander summarizing the total training period shall be submitted to the Chief of Police for review. This report must recommend:
      (1) Retention, based on satisfactory performance, and assignment to a Basic Academy,
      (2) Retention, requiring further training and evaluation prior to assignment to a Basic Academy,
      (3) Termination due to unsatisfactory performance.

IV. BASIC ACADEMY TRAINING
   A. This program of training is designed to provide a detailed, comprehensive program of instruction in the broad field of law enforcement. Its aim is to provide the training essential to enable a police officer to adequately perform routine police duties. This program will cover a two and one-half to five-month period of academic training at a "Basic" Academy approved by the Commission on Peace Officer Standards and Training.
      1. The recruit, although on detached duty during this assignment, will be responsible for compliance to departmental regulations and, additionally, to any regulations specified by the training academy.
      2. It is expected that recruits will demonstrate a high degree of proficiency and achievement in this training program, due to its being their full-time assignment for the duration of the course.

a. A grade average level below "B" may be a contributive factor in denying permanent status.
b. Trainee Evaluation Reports submitted to the department by the academy staff may be influential in determining permanent status.

V. PRACTICAL FIELD TRAINING
   A. This phase of the program is designed to serve as a final evaluative process of training achievement prior to unaccompanied field assignment.
      1. The duration of this assignment is determined by the rapidity of learning and degree of retention of the individual recruit.
      2. The recruit is assigned to field duty as the *active* partner in a patrol unit, under the guidance of a training officer. In this program the recruit will personally perform the required operations demanded by each incident. The training officer will advise and assist as necessary, only to further the training process.
      3. The training officer will instruct and demonstrate for the recruit those procedures, tactics or operations not adequately covered elsewhere in the training program. The training officer will complete the "Training Officer's Checklist" during this phase of the program.
      4. When the recruit demonstrates an ability to satisfactorily perform the routine duties of a police officer without assistance, the training officer will advise the Uniform Division Commander.
      5. If the recruit is unable to demonstrate an ability to satisfactorily perform the routine duties of a police officer without assistance within three months from graduation from a "Basic" Academy, his or her case shall be immediately reviewed to determine the cause.

VI. FINAL REVIEW
   A. After certification by the Uniform Division Commander that direct training has been completed, the recruit shall be assigned to duty as an individual patrol officer. He or she shall be assigned an area of responsibility and shall be held accountable for that area of responsibility.
      1. During this phase of the probationary period, a final determination of the recruit's capabilities should be made in order to recommend retention or termination. Such a determination should be based upon, but not limited to, a demonstration by the recruit of:
         a. Willingness and ability to do the job required.
         b. Ability to apply the knowledge gained from training programs.
         c. Proficiency in performing tasks.
         d. Initiative and enthusiasm.
         e. Firearms proficiency.

(1) Must shoot a qualifying score 8 out of 12 months of the probationary year.
(2) Receives a double issue of ammunition for the first three months of service.
f. Overall rating of Evaluation Reports.
g. Vehicle operation proficiency.
(1) Completion of High-Speed Defensive Driving training course.
(2) General vehicle operation record.
h. Safety record.
(1) The second major "at fault" accident or third minor "at fault" accident can be cause for dismissal.
i. Record of conduct.
(1) The execution of disciplinary actions outside those considered to be advisory, conclusive, or informative are cause for staff review.
j. Compliance with the criteria specified in Section I.

2. Extension of Probation Period

The probationary status of an employee may be extended for six-month intervals for a total period of one additional year. Each recommendation of extension of the probation period must be predicated upon some verifiable, practical reason and be authorized by the Chief of Police.

3. Termination

If a decision is rendered at any time during the probationary period that the recruit is unsuitable for further police service, a Probationary Report is to be completed by the Uniform Division Commander recommending discharge. This report and supporting documents are to be directed to the Chief of Police for review.

4. Retention

Upon completion of the probation period, if the recruit is deemed suitable for police service a Probationary Report is to be completed by the Uniform Division Commander recommending permanent status.

Appendix E

State of New York Municipal Police Training Council

Supervisory Training Course (70 Hours)

1. INTRODUCTION TO THE COURSE .... 1 hour
   Objectives:
   1. To introduce the police supervisor to the new and complex ideas he or she will encounter as a supervisor.
   2. To develop an appreciation of good supervisory techniques.
   3. To provide a basis for the concepts to be developed in the balance of the course.
2. DUTIES AND RESPONSIBILITIES .... 1 hour
   Objectives:
   1. To develop an appreciation of the complexity of the duties and responsibilities of the police supervisor in a modern agency.
   2. To provide a knowledge of the factors involved in defining supervisory responsibilities.
   3. To define the duties and responsibilities of the police supervisor.
3. THE SUPERVISOR AND MANAGEMENT .... 2 hours
   Objectives:
   1. To provide a knowledge of the supervisor's role in management.
   2. To develop an appreciation of the importance of good supervision as a function of police management.
   3. To develop the concepts underlying the supervisory role in police management.

4. PRINCIPLES OF ORGANIZATION .... 2 hours
   Objectives:
   1. To develop an understanding of the importance of good organization to the efficiency of a police agency.
   2. To outline some of the accepted principles of police department organization.

5. DECISION MAKING .... 1 hour
   Objectives:
   1. To develop the ability to make rational decisions using facts and sound judgment.
   2. To provide a knowledge of the factors involved in making decisions.
   3. To develop an appreciation of proper methods and techniques utilized in decision making.
   4. To develop skills in decision making.

6. LEADERSHIP .... 2 hours
   Objectives:
   1. To provide a knowledge of leadership principles in police supervision.
   2. To develop an appreciation of the importance of leadership as a function of police management.
   3. To provide a basic concepts and philosophy of principles of leaders.
   4. To develop leadership skills.

7. PSYCHOLOGICAL ASPECTS OF LEADERSHIP .... 3 hours
   Objectives:
   1. To introduce the concepts of group and individual differences.
   2. To develop an understanding of the ways in which human beings differ from one another.
   3. To develop the concept of the many roles of a supervisor made necessary by differences among subordinates, and of the implication of these roles for police management.

8. ROLE OF SUPERVISOR IN TRAINING .... 3 hours
   Objectives:
   1. To provide a knowledge of current trends in police training.
   2. To develop an appreciation of the importance of training in the police service.
   3. To develop skills in performing the supervisory training function.
   4. To provide basic concepts and philosophy of meeting training needs.

9. TEACHING TECHNIQUES .... 2 hours
   Objectives:
   1. To develop an awareness of some of the effective techniques for use when instructing subordinates.
   2. To provide a knowledge of the use of teaching aids, lesson plans, job analysis, etc., when instructing.
10. WORK PERFORMANCE APPRAISAL .... 2 hours
    Objectives:
    1. To develop the ability to handle work performance appraisal effectively.
    2. To provide a knowledge of the factors affecting work performance appraisal.
    3. To develop an appreciation of proper methods and techniques in working with rating systems.
    4. To develop skills in rating subordinates.
11. PRINCIPLES OF COMMUNICATION .... 2 hours
    Objectives:
    1. To develop an appreciation of the importance of effective communication in supervision.
    2. To provide a knowledge of the basic principles of communication.
    3. To develop skills in communication techniques important to the supervisor.
12. PUBLIC SPEAKING .... 8 hours
    Objectives:
    1. To develop an understanding of the importance of all police officers being able to talk to groups of people.
    2. To provide a knowledge of techniques and methods designed to assist the public speaker.
    3. To provide an opportunity for each student to prepare and deliver a talk before the class.
13. MORALE AND DISCIPLINE .... 2 hours
    Objectives:
    1. To introduce the factors that produce or damage high morale.
    2. To develop an appreciation of the importance of high morale and good discipline in the police organization.
    3. To develop the concepts underlying the supervisory role in morale and discipline.
14. MOTIVATING SUBORDINATES .... 1 hour
    Objectives:
    1. To develop the ability to motivate subordinate workers to work at peak efficiency.
    2. To provide knowledge of factors that cause men and women to increase productivity and efficiency.

3. To develop an appreciation of proper methods and techniques of motivating subordinates.
4. To develop skills in working with subordinates.
15. ELEMENTS OF GOOD REPORTING .... 3 hours
    Objectives:
    1. To provide an understanding of the purpose and need for reports.
    2. To outline the principles of good report writing.
    3. To develop an awareness of the supervisor's responsibility for reports submitted by subordinates.
16. THE POLICE AND THE PUBLIC .... 4 hours
    Objectives:
    1. To develop an understanding of the principle that sound public relations is basic to the effectiveness of the police.
    2. To develop an appreciation of the factors that promote or hinder the development of proper public attitude toward the police.
    3. To provide a knowledge of the steps necessary in the formulation and implementation of a public relations program for police.
17. MAKING DUTY ASSIGNMENTS .... 1 hour
    Objectives:
    1. To introduce the police supervisor to his or her responsibilities in making the most effective use of time and personnel.
    2. To develop an appreciation of the importance of proper distribution of time and work.
    3. To develop skills in making duty assignments.
18. SUPERVISION IN PATROL .... 2 hours
    Objectives:
    1. To develop an appreciation of the importance of efficient patrol service to the success of a police department.
    2. To introduce the supervisor to his or her responsibilities for maintaining a high level of patrol efficiency through proper supervision.
    3. To provide a knowledge of the techniques of patrol supervision which have been successfully applied in police organizations.
19. INTERGROUP RELATIONS .... 2 hours
    Objectives:
    1. To provide a study of the police and minority groups, and problems inherent thereto.
    2. To develop an understanding of how to recognize and handle conflict situations in this area.
20. INSPECTION AND CONTROL .... 2 hours
    Objectives:
    1. To develop an appreciation of the importance of planning, directing, and controlling to the attainment of police objectives.

2. To develop an understanding of the methods used in inspection and control of police operations.
21. CONFERENCE LEADERSHIP .... 8 hours
    Objectives:
    1. To outline the purpose of conferences and the various types of conferences.
    2. To provide an understanding of the principles, procedures, and techniques of handling a conference.
    3. To provide each student with an opportunity to practice leading a conference.
22. TRENDS IN LAW ENFORCEMENT .... 5 hours
    Objectives:
    1. To provide an analysis of present and future problems and prospects in law enforcement.
    2. To develop an appreciation of how good supervision and leadership is necessary to meeting the problems of the future.
    3. To discuss and clarify recent higher court decisions in criminal cases and their effect on police procedures.
23. HANDLING OF UNUSUAL OCCURRENCES .... 2 hours
    Objectives:
    1. To provide a knowledge of the areas of police responsibility in the handling of unusual occurrences in a police jurisdiction.
    2. To develop an appreciation of proper methods and techniques in handling emergencies or disasters.
24. NEWS MEDIA .... 2 hours
    Objectives:
    1. To discuss the area of relations between the police and the news media, and problems related thereto.
    2. To develop guidelines and standards for the dissemination of police information to the news media.
25. CASE STUDIES OF SUPERVISORY PROBLEMS .... 2 hours
    Objectives:
    To provide an opportunity for application of supervisory techniques to typical problems through the study of case problems, both actual and hypothetical.
26. EXAMINATION AND REVIEW .... 2 hours
    Objectives:
    1. To test by objective questioning the students' understanding of the material presented.
    2. To review with the students the examination questions to clear any misunderstandings.
27. CRITIQUE OF COURSE (Panel Q & A) .... 2 hours
28. GRADUATION .... 1 hour

Appendix F

**California Program for Peace Officers' Training
Qualities of Personal Leadership—A Supervisor's Rating Scale**

1. Forcefulness
   a. Do I give orders properly and see that they are followed out, maintaining a businesslike attitude constantly?
   b. Do I keep in touch with the efforts of my subordinates so that I know how well each is doing?
   c. Do I preserve the right balance between too much sternness and too much familiarity?
2. Ability to inspire confidence
   a. Do I show respect for my subordinates and for myself?
   b. Am I impartial, or do I play favorites?
   c. Do I exercise self-control, or do I allow my temper frequently to get the better of me?
3. Ability to take a personal interest in subordinates
   a. Do I talk with subordinates as people, rather than as inferiors?
   b. Do I give them personal training and discuss their work with them?
   c. Do I get things for them which they would be unable to get without my assistance?
   d. Do I help them to realize their ambitions?
4. Ability to get the work done correctly
   a. Do I give instructions so clearly that no one can misunderstand what is wanted?
   b. Do I check up on my men and women to see that my orders are followed out exactly?

5. Ability to get and use the ideas of men and women
   a. Am I successful in getting suggestions from subordinates?
   b. Do I use these suggestions when I get them?
   c. Do I give credit to the man or woman who gives me an idea when I am talking about it to my superiors and colleagues?
6. Ability to work with people rather than over them
7. Ability to lead rather than boss
   a. Do I show the men and women how they can work more efficiently, rather than ordering them about without showing them how?
   b. Do I train them in better methods?
   c. Do I set the example by being as hard on myself as I am on any of my subordinates?
8. Ability to develop teamwork
   a. Am I careful to plan ahead?
   b. Is the mechanical equipment for which I am responsible always ready for work?
   c. Do I place the right people in the right positions?
   d. Do I allocate responsibility for results so that my people know what they have to do?
   e. Does the spirit of teamwork exist among my people?
9. Ability to show kindness without being considered easy
   a. Do I remember that my people are human beings and treat them with common courtesy?
   b. Do I work for the interest of my people?
   c. Do I know how to keep the men and women from imposing on my good nature?
   d. Can I properly balance praise and censure?
10. Ability to reprimand properly
    a. Do I always make sure of my case before I reprimand?
    b. Do I give reprimands in private, except in unusual cases?
    c. Do I reprimand in a straightforward manner, or do I merely nag?
    d. Do I give the reasons for my reprimands?
    e. Do I follow up the reprimands?
    f. Can I reprimand good people and make them feel that it is fair?
11. Ability to keep from worrying
    a. Do I entrust responsibility to my men and women, allowing them to make some mistakes?
    b. Do I train them on the job so that they can take over work that I ought to give them?
    c. Am I willing to delegate work, or do I feel that I want to do everything myself?
12. Ability to call forth the best efforts of the men and women
    a. Can I develop enthusiasm in my people?
    b. Do I know how to prevent idling and carelessness?

13. Ability to train men and women on the job
    a. Do I know how to analyze a job before teaching it to a beginner?
    b. Do I show the person carefully how to do it?
    c. Do I let the person try it while I watch?
    d. Do I correct the person's mistakes?
    e. Am I in the habit of keeping my eye on a beginner until he or she is able to do the job well?
14. Ability to make a new officer feel at home
    a. Do I introduce the new officer to the older personnel?
    b. Do I show a personal interest in the new officer?
    c. Do I make it easy for the new officer to ask questions?
15. Self-confidence
    a. Am I sure of myself on the job, or am I afraid of it?
    b. Do I help my subordinates to overcome self-consciousness?
    c. Do I show subordinates that they are better than they think they are?
16. Accident prevention
    a. Do I believe in safety and accident prevention?
    b. Do I know that most of the men and women working for me have never had to do the kind of work they are doing today and have not heard of a safety program?
    c. Do I wait for an accident to happen before correcting an unsafe condition—or do I plan ahead?
    d. Do I read safety bulletins and explain them?
    e. Do I just *tell* my people to "be careful," or do I *show* them the safe way and why?
    f. Do I assume the responsibility for safety conditions on my job, or do I "pass the buck" along to someone else?

Appendix G

**California Program for Peace Officers' Training
A Form for Rating Supervisors**

Report on Performance of _____ Made by _____

Department _____ Title _____

Directions to the officer making the report: Place an "X" in the blank space after each item on this and the following page which describes the work, performance, and conduct of the Supervisor, whose name appears above, for the past two months. Check only one item in each of the groups numbered 1 to 19. Check as many items as apply in groups 20 and 22. Write in any needed information in groups 21 and 23.

1. Relations with other supervisors:
   a. Often not satisfactory _____
   b. Sometimes not satisfactory _____
   c. Usually gets along well _____
   d. More satisfactory than average _____
   e. Exceptionally satisfactory _____

2. Knowledge of the characteristics and ability of subordinates:
   a. Knowledge markedly limited _____
   b. Knowledge somewhat limited _____
   c. Knows employees fairly well _____
   d. Knowledge better than average _____
   e. Knowledge exceptional _____

3. Skill in training subordinates:
   a. Often ineffective or limited _____
   b. Somewhat limited or ineffective _____
   c. Fairly effective _____
   d. More effective than average _____
   e. Exceptionally effective _____

4. Fairness and impartiality in dealing with subordinates:
   a. Often unfair or harsh _____
   b. Sometimes unfair or harsh _____
   c. Fairly just _____
   d. Better than average _____
   e. Exceptionally fair and just _____

5. Character of discipline:
   a. Often ineffective or inadequate _____
   b. Sometimes ineffective or inadequate _____
   c. Fairly suitable and effective _____
   d. More effective than average _____
   e. Exceptionally effective _____

6. Willingness to make difficult decisions:
   a. Often "passes the buck" _____
   b. Inclined to "pass the buck" _____
   c. Usually properly willing _____
   d. More willing than average _____
   e. Exceptionally willing _____

7. Success in making and carrying out plans in an orderly manner:
   a. Often not successful _____
   b. Sometimes not successful _____
   c. Usually fairly successful _____
   d. Better than average _____
   e. Exceptionally successful _____

8. Knowledge of the work supervised:
   a. Knowledge markedly limited _____
   b. Knowledge somewhat limited _____
   c. Knows work fairly well _____
   d. Better informed than average _____
   e. Exceptional knowledge of work _____

A FORM FOR RATING SUPERVISORS

9. Willingness to take on new or additional work:
   a. Often too reluctant or unwilling _____
   b. Sometimes too reluctant or unwilling _____
   c. Usually properly willing _____
   d. Better than average _____
   e. Exceptionally willing _____

10. Resourcefulness in meeting difficulties:
    a. Often goes to pieces _____
    b. Usually meets the situation _____
    c. Easily discouraged by obstacles _____
    d. More resourceful than average _____
    e. Nearly always finds good way out _____

11. Skill in giving and following up orders and assignments:
    a. Often vague or inadequate _____
    b. Somewhat vague or inadequate _____
    c. Fairly definite and adequate _____
    d. Better than average _____

12. Ability to learn new work:
    a. Learns with difficulty _____
    b. Learns somewhat slowly _____
    c. Learns fairly easily _____
    d. Better than average _____
    e. Learns with exceptional ease and speed _____

13. Initiative in own work:
    a. Often lacking in initiative _____
    b. Sometimes lacking in initiative _____
    c. Shows fair initiative _____
    d. More than average _____
    e. Shows exceptional initiative _____

14. Control of temper and emotions:
    a. Emotions often impair work _____
    b. Emotions somewhat impair work _____
    c. Emotions usually well controlled _____
    d. Control better than average _____

15. Loyalty to job and employer:
    a. Often not loyal _____
    b. Sometimes not loyal _____
    c. Usually loyal _____
    d. More loyal than average _____
    e. Exceptionally loyal _____

16. Quality of work done by subordinates:
    a. Faults frequent or serious _____
    b. Work sometimes careless _____
    c. Work usually free from serious or numerous faults _____
    d. Quality better than average _____
    e. Work nearly always free from faults _____

17. Volume of work done by subordinates:
    a. Markedly limited _____
    b. Somewhat limited _____
    c. Average for the work _____
    d. More than average _____
    e. Exceptional _____

18. Working and directing others:
    a. Often does too little or too much of work _____
    b. Sometimes works or "bosses" too little or too much _____
    c. Satisfactory combination of working and directing _____
    d. Better combination than average _____
    e. Nearly always combines working and "bossing" exceptionally well _____

19. Standing in own group and field as a supervisor:
    a. Near the bottom—inexpert _____
    b. Below the average—indifferent _____
    c. Average—fairly competent _____
    d. Above the average—very good _____
    e. Near the top—exceptionally good _____

20. Miscellaneous favorable traits: (Check such of these items as apply and as are advantageous in the work.)
    a. Has helpful outside contacts or interests _____
    b. Well trained technically for the work _____
    c. Often substitutes in higher positions _____
    d. Working hard to fit self for advancement _____
    e. Exceptionally courteous to others _____
    f. Has unusual "common sense" _____
    g. Has fine sense of humor _____
    h. Exceptionally good appearance _____
    i. Has unusual interest in work _____
    j. Welcomes suggestions from superiors _____
    k. Uses exceptionally good English _____
    l. Observes safety practices closely _____

21. Other favorable items: (specify)
    a. _____
    b. _____

22. Miscellaneous unfavorable traits: (Check such of these items as apply and affect the work adversely.)
    a. Unduly opinionated or stubborn _____
    b. Given to sarcasm _____
    c. Inclined to gossip _____
    d. Wastes much mental or physical effort _____
    e. Lacks "common sense" _____
    f. Outside interests impair work _____
    g. Lacks proper personality for a supervisor _____
    h. Has poor sense of humor _____
    i. Careless about appearance _____
    j. Uses markedly poor English _____
    k. Lacks needed technical training for work _____
    l. Negligent in safety matters _____
    m. Supervision markedly "hardboiled" _____
    n. Supervision lax _____
    o. Too familiar with workers _____

23. Other unfavorable items: (specify)
    a. _____
    b. _____

Appendix H

California Program for Peace Officers' Training Self-Analysis Questionnaire for Supervisors

Directions: Check the reply which characterizes your *usual* action or attitude.

                                                                                                Yes   No

1. *Health and energy:*
    a. Do I take care of my health so as to be physically fit for my job?
    b. Do I worry too much about myself, my home, or my job?
    c. Do I find myself excessively fatigued during working hours?
    d. Am I often "crabby" to my subordinates because I don't feel well?
    e. Am I often ill, necessitating absence from work?
    f. Do I direct my energy where it will be most productive?

2. *Drive:*
    a. Am I conscious of an inner urge which carries me forward in my work?
    b. Am I able to transfer this urge to my staff?

3. *Ambition:*
    a. Do I believe in the worthwhileness of the work of the Division?

SELF-ANALYSIS QUESTIONNAIRE FOR SUPERVISORS

   b. Can I see the possibility of personal growth?
   c. Does my personal goal coincide with the objectives of the Division?
4. *Perseverance:*
   Do I attempt another method when the first has failed?
5. *Ability to Reach a Decision:*
   a. When matters come before me for decision, do I decide them with reasonable promptness?
   b. Do I usually feel that I have at hand enough information and knowledge on which to base my decision?
6. *Courage of one's convictions and the ability to carry responsibility:*
   a. Once I have made a decision, have I the ability to put it out of my mind and not worry about it?
   b. Am I willing to assume the responsibility not only for my own acts but for those of my subordinates?
7. *Open-mindedness and receptiveness:*
   a. Am I successful in getting suggestions from my staff?
   b. Do I use these suggestions when I get them?
   c. Do I give credit to the man or woman who gives me an idea when I am talking to my supervisors and colleagues?
   d. Am I able to put my subordinates at ease when they come to me to discuss some matter?
   e. Am I conscious of the fact that I have some prejudices, and am I on my guard against them?
   f. Do I make it easy for a new worker to ask questions?
   g. Am I big enough to change my decision and accept the responsibility?
8. *Fairness:*
   a. Do I show respect for my staff and myself?
   b. Do I show evidence of favoritsm?
   c. Can I reprimand a good worker when necessary and make him or her feel that it is fair?
   d. Am I so obviously unbiased that no one can suspect any prejudice in my official relationships?
9. *Firmness:*
   a. Do I insist that all orders be carried out or else countermand them?
   b. Do I know how to keep my people from imposing on my good nature?
   c. Am I able to draw the line between firmness and obstinacy?

10. *Cheerfulness:*
    a. Do I permit my outside problems or difficulties to influence my disposition during working hours? ___ ___
    b. Can I maintain a cheerful demeanor when things go wrong? ___ ___
    c. Can I be serious when the occasion warrants it without becoming gloomy and pessimistic? ___ ___
11. *Enthusiasm:*
    a. Can I develop enthusiasm in my subordinates? ___ ___
    b. Do I know how to prevent idling and carelessness? ___ ___
    c. Can I make a new worker enthusiastic about his or her new job? ___ ___
    d. Can I stimulate enthusiasm when it is most needed to buoy up my staff as well as myself? ___ ___
12. *Self-control and calmness:*
    a. Do I exercise self-control, not permitting my temper to get the better or me? ___ ___
    b. Can I remain calm in emergencies? ___ ___
    c. Can I handle angry, rebellious, or sarcastic individuals without using the same manner? ___ ___
13. *Consistency:*
    a. Do I frequently give one set of orders and then immediately change them? ___ ___
    b. Do my emotions interfere with my work to such an extent that my staff does not know what to expect? ___ ___
    c. Do I possess an evenness of disposition? ___ ___
14. *Industry:*
    a. Do I plan my work so that my expenditure of time and energy will be as productive as possible? ___ ___
    b. Do I do my share of the work? ___ ___
    c. Do I get my own work finished on time? ___ ___
15. *Forcefulness:*
    a. Do I give my orders properly and see that they are carried out, maintaining a businesslike attitude constantly? ___ ___
    b. Do I keep in touch with the efforts of my staff so that I know how well each one is working? ___ ___
    c. Do I preserve the right balance between too much sternness and too much familiarity? ___ ___
16. *Initiative:*
    a. Am I able to discern the need, when it arises, for initiating action? ___ ___
    b. Do I know what action is within my province to initiate and what should be first discussed with my superior? ___ ___

c. Have I the ability to plan the necessary action?
17. *Technical knowledge:*
    a. Do I understand thoroughly how each step of the work is to be accomplished?
    b. Do I know the function of my part of the work in relation to the accomplishment of the whole?
    c. Do I know what the acceptable standards are for the work I supervise?
18. *Organizing ability:*
    a. Do I understand precisely of what my own responsibilities consist?
    b. Am I willing to delegate work instead of feeling that all work must be done by myself?
    c. Do I try to plan my work ahead?
    d. Do I find myself unduly occupied with details?
    e. Do my plans remain merely plans?
    f. Does the control of the operation for which I am responsible actually center in me?
    g. Have I in hand at all times all essential information for directing the operation?
    h. Do I give my associates and subordinates a sense of participation or partnership in planning and conducting the operation?
    i. Do I thrust responsibility on subordinates who are not ready for responsibility, thus allowing them to make mistakes?
19. *Frankness:*
    a. Do I use the same frankness with my men and women as I expect from them?
    b. Do I discuss with them any personal shortcomings which are detrimental to them?
    c. Do I temper my frankness with tact?
20. *Tact:*
    a. Do I always make sure of my case before I reprimand?
    b. Do I give reprimands in private?
    c. Can I properly balance praise and censure?
    d. Do I give reasons for my reprimands?
    e. Do I use care in handling difficult situations in order not to offend or hurt my subordinates?
    f. Do I avoid saying things which will leave the worker sullen or antagonistic?
21. *Simplicity and clarity:*
    a. Do I give instructions so clearly that no one can misunderstand what is wanted?

    b. Do subordinates have a clear idea of their responsibility? _____ _____
    c. Do I adapt my vocabulary to the individual I am addressing to insure that it is perfectly clear? _____ _____
    d. Do I become impatient when it is necessary to repeat explanations because they are not understood? _____ _____
22. *Teaching skill:*
    a. Do I show the men and women how they can work more efficiently rather than order them about without showing them how? _____ _____
    b. Do I train them in better methods? _____ _____
    c. Do I know how to analyze a job before teaching it to a beginner? _____ _____
    d. Do I show the person carefully how to do it? _____ _____
    e. Do I let the person try it while I watch? _____ _____
    f. Do I correct mistakes? _____ _____
    g. Am I in the habit of keeping my eye on a beginner until he or she is able to do the job well? _____ _____
    h. Do I train my people on the job so that they can take over work that I ought to give them? _____ _____
23. *Ability to evaluate capacities of staff:*
    a. Am I able to determine what each member of my staff is capable of doing? _____ _____
    b. Can I determine whether each staff member is doing the best work of which he or she is capable? _____ _____
24. *Social-mindedness:*
    a. Do I understand the social and economic problems of the groups with whom I come in contact? _____ _____
25. *Ability to cooperate:*
    a. Does the spirit of teamwork exist among my staff? _____ _____
    b. Do I place the right person in the right position? _____ _____
    c. Do I work with them rather than over them? _____ _____
    d. Do I carry out orders I receive in the same manner as I expect my subordinates to carry them out? _____ _____

Appendix I

**Berkeley Police Department
Field Training Guide**

Date Received	Officer

Date Completed	Instructor

USE OF THE FIELD TRAINING GUIDE:

The Field Training Guide serves two basic purposes: (1) It is a guide for the instructor to follow in training a new officer. (2) It prevents unnecessary duplication in training where there is more than one instructor.

The items of instruction listed in the following pages include basic jobs and necessary information that every officer should know.

When the item has been explained and demonstrated, wherever possible the instructor shall place his initials opposite the item.

When the Field Training Guide has been completed, the Instructor shall return it to the Personnel Officer.

I have been instructed in all items listed in the Field Training Guide, and have completed the basic reading assignments.

Signed_____

Date & Time_____

_____ 1. Introduction to other officers.
_____ 2. Tour of the Police Department.

_____ 3. *Orientation:* Police Department; city departments; related facilities.
_____ 4. *Equipment to be issued:*
Citations
Bicycle tags
Day book paper and dividers
Information sheets
Radio code
Radio car assignment card
Traffic code brief
Traffic ordinance
Vehicle code
Bicycle ordinance
City map
Traffic chalk
Printed forms: Invitation to Call, Open Door and Windows, Vacation cards.
_____ 5. General Orders.
_____ 6. Rules and Regulations.
_____ 7. Checking on and off time sheet.
_____ 8. Days off, lunch relief, sick relief, overtime, annual leave, military leave of absence, leave of absence, recovery time, personnel ordinance and PR forms.
_____ 9. Basic shift, overtime, court time, special duty. Reference General Order D-16 and PR appendix 10.
_____ 10. Revolver and its use, range practice and the use of the range, carrying gun and use off-duty. PR 317-324. Submit report to Personnel Captain on service revolver to be used.
_____ 11. Use of baton and handcuffs. PR 314.
_____ 12. Mailbox section: use, restrictions, and need for constant checking.
_____ 13. Public relations: Press policy and availability of information. PR 226-227.
_____ 14. Attorneys, bail bondsmen—policy of no recommendation. PR 212-214.
_____ 15. Departmental policy concerning gratuities and rewards. (not permitted PR 215-217).
_____ 16. Eating, sleeping and study habits re home life.
_____ 17. Personal conduct rules. PR 228-252.
_____ 18. Inspection drill, observation, and demonstration.
_____ 19. Police library, use of, recommended reading.
_____ 20. Basic reading assignment:
    a. Local Ordinances:
        553 NS Noise-nuisances
        1358 NS lb—Drunk in auto; private property.

1912 NS Unlawful entry of theaters.
2016 NS False Police Report.
2130 NS Soliciting contributions.
2345 NS Bike Ordinance.
2532 NS Control of dogs and other animals.
2795 NS Disorderly Conduct.
2800 NS Taxi Ordinance.
2829 NS Trespassing posted property.
2858 NS Solicitors and Peddlers.
2847 NS Distribution of Advertising Matter.
2878 NS Firearms; BB guns, etc.
2881 NS Dangerous weapons.
2899 NS Gambling.
3262 NS Traffic Ordinance.
3504 NS Curfew—Minors on the street.
3602 NS Refuse dumped on City Street.
3289 NS Dance Permits.

b. Penal Code:
Read thoroughly all of Chapter V covering laws of arrest.

Sections:	148	242	286	460	625a
	148.1	243	288	484	647
	148.5	244	288a	487	647a
	187	245	314	499b	650½
	211	246	415	594	
	216	261	417	602	
	217	272	459	622	

Dangerous Weapons Control Law: 12000 thru 12551 PC

c. Code of Civil Procedure: Rules of Evidence; 1823—1839 CCP.
d. Business and Professions Code: Alcoholic Beverage Control Act; 25600 thru 25665.
e. Health and Safety Code. Chapter Five: Narcotics 11500 thru 11557.
f. Welfare and Institutions Code. Sections 600 thru 641.
g. Vehicle Code:
   Division 3, Chapter 1.
   Division 4, Chapters 1 thru 6.
   Division 6, Article 4 of Chapter 1, and Chapter 4.
   Division 9, Chapters 1 and 2.
   Division 10.
   Division 11.
   Division 12, Chapter 1 thru 5.

____ 21. Personal Appearance. PR 223; 300.
____ 22. Daily bulletin, use and general purpose, need for checking bulletin on reporting to work and again prior to leaving.

_____ 23. Squadroom files, information available.
_____ 24. Police responsibilities; protection of life and property; regulatory duties.
_____ 25. Corporation yard. How to draw gasoline and oil, emergency repairs to vehicle, car number, etc.
_____ 26. Siren and red light. Restrictions on its use. Safety precautions when using.
_____ 27. Uniform equipment. Required—optional. PR 300.
_____ 28. Desk operation. Radio communication, assignment, beat organization, police call box system, coordination with re-call signal and radio.
_____ 29. Availability of personnel.
_____ 30. Organization of the Police Department.
_____ 31. Address and telephone change forms.
_____ 32. Sick card, representation card, and recovery requests. How prepared and to whom submitted.
_____ 33. Courtesy. Establish and practice habit of courtesy in all contacts, public, personal, and official.
_____ 34. Sergeant's responsibility to patrol officer—advice, supervision and assistance.
_____ 35. Monthly statistical report. Correlate with probation report.
_____ 36. Fleet vehicles, care and use of equipment therein. Never leave auto interior dirty.
    a. Check prior to use.
    b. Check after use.
    c. Never leave low on fuel.
    d. How to report defective condition.
_____ 37. Personal safety—Necessity for examination and treatment by doctor from Industrial Medical Group for service-connected injuries.

Routine Matters

_____ 1. How to use Police Call Box.
_____ 2. How to check off and back from eating.
_____ 3. How to check off and back from beat, coffee, necessity for remaining available.
_____ 4. How to make an hourly ring.
_____ 5. Police lights, their use, location, acknowledge by radio or phone box.
_____ 6. How to keep desk officer informed of your status, availability, and checking back with desk after an emergency.
_____ 7. Organization and use of the day book.
_____ 8. Geographical features, streets, block numbers, shortest routes, dead ends, schools, buildings, fire houses, parks, transportation

FIELD TRAINING GUIDE   289

facilities, beat boundaries, policy boxes, signal lights, escape routes from various areas.
_____ 9. How to make self conspicuous as crime preventive measure and inconspicuous at other times. Why this is necessary.
_____ 10. How to recognize a series of crimes (MO repetition). Procedures for combating crime series.
_____ 11. How to get acquainted with merchants, taxi drivers, delivery people, cleanup people, etc. Observation of routine habits and hours of employment.
_____ 12. How to suggest security measures to store owners and people in business. Alley lights, safe lights, night lights. Better locks, barring windows, alarm systems. Placing safe where can be seen, lighting and security. Recommending visit or assistance of inspector.
_____ 13. How to recognize an "Attractive Nuisance" and other crime hazards. How to reduce such hazards.
_____ 14. How to furnish information to citizens and where to get information if you don't know the answer.
_____ 15. How to check hotels for wanted and missing persons. Availability of register (1846 N.S.).
_____ 16. How to develop and use sources of information.
_____ 17. How to check autos and bicycles against wanted list (hot file).
_____ 18. How to report broken water mains, PG&E or PT&T lines, street lights and electroliers, signal lights, night lights, open doors and windows, defective sidewalks and streets. How to detect responsibility and report to the desk. Use of defective condition form on city property.
_____ 19. How to observe and handle unhealthy conditions:
   a. Dead or disabled animals.
   b. Improper garbage disposal.
   c. Improper disposal of debris.
   d. Fire hazards.
   e. Health and safety hazards.
_____ 20. How to check schools, playgrounds, parks, and recreation centers for sex offenders and loiterers.
_____ 21. How to investigate peddlers and solicitors (police permit).
_____ 22. How to make daily bank inspections. (21 and 22 usually occur in daytime but First Platoon officers should understand because they do work days occasionally during football season.)

Police Tactics

_____ 1. Use of auto:
   a. Driving and parking on routine patrol.

b. Driving and parking in emergencies.
   What constitutes an emergency.
   Efficient driving and parking habits.
   Defensive driving.
c. Prowler, alarm, robbery, or burglary in progress. How to approach.
d. Quadrant covering.
e. Fixed post in auto or on foot.
   How to arrive at post.
   What route to take in approach.
   Use of red light and siren.
   Danger of too rapid approach.
f. Traffic enforcement.
   Red light and siren.
   Following the violator—safety—stopping.
   Parking.
   Leaving scene after stop, turning off.
g. Answering fire calls.
   Speed of approach, mindful of hydrants.

_____ 2. How to approach scene on foot on routine assignments and emergencies.
   Quadrant search.
_____ 3. How to request assistance of other officers.
_____ 4. How to proceed when answering a police call.
_____ 5. How to proceed when assigned to a report of a burglary, sex offense, or robbery when person responsible has left the scene.
_____ 6. How to guard or transport money.
_____ 7. How and where to check for vagrants.
_____ 8. How to proceed when assigned to fire. Duties at scene.
   Examples:
      a. Arson, etc.
      b. False fire alarms.
      c. Protecting fire lines and hoses.
      d. Guarding fire equipment.
      e. Second alarm fires requiring ambulance service
         (three alarm in residential areas).
_____ 9. How to check a store group by auto and on foot.
   a. Quiet and alert—keys in pocket.
   b. Use of flashlight, gun.
   c. Observant for attacks on store, past and fresh, insecure locks, lights out, open doors or windows and reporting same. Necessity of marking past attacks.
   d. Inspecting alleys, rooftops, rear yards, places of concealment, methods of approach and exit.

FIELD TRAINING GUIDE   291

        e. Making game out of tedious job. Erratic checking, skip checking, watch and wait.
        f. Calling owners if store open, notifying desk of situation. Store information.
        g. Use of imagination—assumed conditions as practice method.
____ 10. Suspicion. How to be suspicious. Persons observed and questioned.
____ 11. Stop Cards:
        a. Recording names, places, times, and dates.
        b. Value to their use.
        c. Distribution of copies.

Miscellaneous Assignments

____ 1. How to secure and dispose of garage tags.
____ 2. Flags, where staffed, normal care of, storage when wet.
____ 3. How to take out and bring in mail.
____ 4. How to deliver copies of "Hot Sheet" to adjoining jurisdiction.
____ 5. How to check university files for student addresses.
____ 6. How and where to pick up copies of the Oakland Police Bulletin.
____ 7. Where to pick up "Daily Call" and routing of same.
____ 8. How to guard a prisoner at the hospital, elsewhere.
____ 9. How to check a vacation house:
        a. Reason for vacation house service.
        b. Method of distribution.
        c. Different types, numbered and unnumbered.
        d. Dwellings considered eligible.
        e. Time to check house, daylight hours.
        f. Checking doors and windows, pick up papers.
____ 10. How to transport juveniles to the Alameda County Detention Home.
        a. Who must go, forms to accompany.
        b. Location and procedure on arrival.
        c. Refer to General Order J-18.1, arrest of juveniles.
____ 11. How to drive ambulances and patrol wagon, location of equipment and its use. Refer to G.O. E-1, E-2, and E-4.

Investigative Principles

____ 1. How to handle a chronic complainant. Procedure in such cases.
____ 2. How to determine if complaint is unjustified.
        a. Unfounded report. Refer to G.O. R-16.
        b. False police report. Refer to 2016 N.S.

3. How to determine if crime has been committed.
     a. Aids in determining responsible person.
  4. How to protect a crime scene.
     a. How to collect physical evidence.
  5. How to prepare investigation notes for noninjury and injury collisions.
     a. Notes attached to investigation report.
     b. Preservation of notes (Court appearances, etc.).
     c. Rough sketch necessary.
  6. Descriptions: general—peculiarities—outstanding features—identification value.
  7. How and when to diagram crime scene. Photographs and other procedures.
  8. How to interview persons. Rooms to use at the Hall of Justice. Interviews at other locations, value of written statements. Stress G.O. H-2 re mentally ill and drunks in interview rooms.
  9. How to interrogate suspects: Interrogation rooms, presence of friends and relatives. Methods: promises, fear, or coercion not permitted. Admission and confession defined. Admonitions.
  10. How to search files: record bureau, outside departments, S.I.B. and Security, crime files.
  11. How to confer with inspectors.
  12. Procedures for identifying suspects:
     a. Photographs
     b. Lineup
  13. Evidence: How to mark, transport, preserve, wrap, label, and place in property room. Must prevent contamination and maintain chain of evidence. See G.O. P-61 and P-65.
  14. How to report a vice condition. Steps to take and to whom information forwarded.
     a. Narcotics
     b. Gambling
     c. Prostitution
     d. Liquor violations

Persons

  1. What to do with a found person.
     a. Juvenile
     b. Confused adult
     c. Senile or mentally confused
  2. Missing person reports.
     a. Juvenile under 12 years of age, necessity of continuing investigation.

FIELD TRAINING GUIDE   293

        b. Adult. Consider possibility of: crime, marital difficulties, home problems, whether missing or not.
        c. Forms to be filled out. Case in pending status until subject located.

_____ 3. How to investigate casualty reports, including miscellaneous accidents, seizures, and emergency sick persons.

_____ 4. Preliminary investigation of dead body cases.
        a. Determination of cause. Natural vs. unknown cause.
        b. Disposition of body, coroner's case or death certificate by physician.
        c. Need for complete details with CC for DD to act as coordinator with County Coroner at the Coroner's hearing.
        d. General Order O-1.

_____ 5. How to proceed with a mentally ill subject.
        a. When commitment ordered by Superior Court.
        b. Police responsibility.
        c. Emergency commitment to Alameda County Hospital; legal requirements and observations.
        d. Custodial care and precautions to prevent injury to self and subject.
        e. General Order I-16.

_____ 6. How to care for unconscious persons.
        a. Prisoners not to be placed in City Jail in unconscious condition. Take to HMH for observation.
        b. Notification of relatives or friends. Care of and transportation to HMH for medical treatment.
        c. Identification of subject same as for found person.

_____ 7. Basic First Aid.

_____ 8. How to recognize and arrest vagrants, drunks, drug addicts and disorderly persons. Consider possibility of illness, injury, mental defect, diabetic coma and insulin shock.

_____ 9. Arrest of Armed Forces personnel, disposition, drunk, AWOL.

Citizens' Arrests
        a. Legal obligation if citizen wishes to make an arrest.
        b. Procedure for obtaining a citizen's complaint.
        c. Importance of signed statements.
        d. Prisoner to be released if complaint not signed on first Court day following arrest. To be so indicated on arrest card.
        e. Distinction between felony and misdemeanor arrest procedure.
        f. Provisions of General Order A-50.

Juveniles

_____ 1. Juvenile Procedures (General Orders J-16, J-17.3, J-18.2, J-3.2.)
   a. Minor is any person under age of 21.
   b. General Orders define juvenile as any person under 18 years of age.
   c. Parents to be notified in all cases where juveniles detained, whether actual arrest made or not.

_____ 2. How to approach and talk to juveniles.
   a. Juvenile competency must be evaluated.
   b. Clearance of juvenile crimes.
   c. Crimes seldom committed alone.
   d. Neighborhood juveniles as a source of information.

_____ 3. Juvenile traffic offenders.
   a. No appearance in municipal court.
   b. All under 18 years cited to appear at Juvenile Bureau within 5 days.

_____ 4. Juvenile intoxicated or in possession of intoxicating beverages (Alcohol Beverage Control Act).
   a. Age in this instance is 21 years.
   b. Read 25658, 25660.5, 25661, 25662, 25665. Covers liquor in possession, false I.D., and licensed premises.

_____ 5. How to file a Juvenile Responsibility Report.
   a. "Notice to Parent" form when juvenile released.
   b. Juvenile loitering form.

Animals

_____ 1. How to place an animal in the pound.
   a. Where to place.
   b. Notify City Pound.

_____ 2. How to dispose of dead animals.
   a. Remove to City Pound—notify.
   b. If removed from sight, a note to Pound indicating location and requesting them to pick up.

_____ 3. How to handle cases of lost, found, and wounded animals.
   a. Checking animal license at City Hall.
   b. Wounded animals to Pet Hospital if Pound closed.
   c. Copies of report to Pound.
   d. General Order A-32.

_____ 4. How to handle a report of an animal bite.
   a. Forms to be used—routing.
   b. Locating responsible animal.
   c. Where animal can be impounded, exceptions.
   d. Follow-up on case not closed on first contact.
   e. General Order A-31.

Traffic

_____ 1. Vehicle Code.
   a. Pertinent sections.
   b. Necessity for reading and understanding balance of code.
   c. Recognizing a traffic violation.
   d. Issuing the traffic citation.
   e. When officer must arrest at scene.
   f. When officer has option of arrest or citation.
   g. Complaint and warrant for later arrest.
_____ 2. How to direct traffic.
   a. At an accident scene.
   b. At a street corner without signals.
   c. At a street corner with signals; manual control.
   d. General rules of traffic control.
   e. Control no longer than necessary.
   f. Taking post where clearly visible.
_____ 3. How to handle an accident.
   a. Noninjury.
   b. Injury.
   c. Care of injured.
   d. Determination of seriousness of accident.
   e. Locating drivers.
   f. Clearing scene to allow free flow of traffic. Removing glass, gasoline, BFD.
   g. Statements from drivers and witnesses.
   h. Photographs, diagrams, and measurements.
   i. How to determine responsibility—citation.
   j. General Order T-16.
_____ 4. How to recognize and proceed with a drunk driver.
   a. Observations.
   b. Blood alcohol test—significance of BA reading.
   c. Physical condition sheet.
   d. Operation of tape recorder and value of taped interview.
_____ 5. How to investigate a hit-run accident.
   a. Classification of offense depends upon injury to persons other than driver.
   b. Physical evidence left at scene.
   c. Location of witnesses.
   d. Possibility of following trail of evidence to responsible party.
_____ 6. How to recognize traffic violators on routine patrol.
   a. Application of principle of selective enforcement.
   b. Public acceptance.
   c. Use of traffic charts, reports, and spot maps in traffic bureau office.
_____ 7. How to issue a citation.

        a. Citing to court office; when.
        b. Citing to court; when.
_____ 8. How to handle a misdemeanor Hit and Run case.
_____ 9. Vehicle Code violations handled by citations but requiring numbered cases.
     23109VC
     23122VC
     23123VC
     23103VC—when no arrest is made.
_____ 10. How to handle a 14601VC complaint after information returned from DMV.

Arrest Mechanics

_____ 1. How to search a male.
        a. At scene.
        b. At station.
_____ 2. How to search a female.
        a. At scene.
        b. At station.
_____ 3. How to make an arrest.
        a. Distinguish between felony and misdemeanor arrest procedure.
        b. Use of force—matter of law—departmental policy.
        c. Resisting arrest.
        d. Precautions with prisoner.
        e. When to handcuff.
_____ 4. How to transport prisoners to the station.
        a. Use of auto; alone, with another officer.
        b. Patrol wagon; when to use, following to station.
        c. Felony prisoners, extra precautions.
        d. Necessity for care and watchfulness, prevent prisoner from getting behind officer.
_____ 5. How to handle arrested persons at station.
        a. Search in jail office—by jailor—witnessed by officer.
        b. Arrest card, different types.
        c. Receipt for property, why three copies.
        d. Posting in arrest book.
        e. Search of files.
        f. Disposition of arrest card.
        g. Signing complaints.
        h. Determining amount of bail.
        i. Arrest of lodgers, how handled.

FIELD TRAINING GUIDE

Police Procedures

Minor Crimes and Civil Matters

_____ 1. Illegal parking: moving cars, towing.
_____ 2. Domestic matters: arbitrating, warning, arresting.
_____ 3. Bonfire: dumping garbage.
_____ 4. Noise complaints: types.
_____ 5. Disorderly conduct and disturbances: adults, juveniles, students.
_____ 6. Civil matters: neighborhood disputes, landlord-tenant affairs.
_____ 7. Malicious mischief.
_____ 8. Trespassing.

Major Crimes Usually Requiring M.O. Crime Reports

_____ 1. Assaults, felonies.
_____ 2. Sex offenses.
_____ 3. Stolen autos: joy riding, signing forms, wanted circular, broadcast, teletype (cancellation when recovered).
_____ 4. Theft cases: grand, petty, purse snatch, bunco, shoplifting, car clouting.
_____ 5. Robbery.
_____ 6. Burglary.

Criminal Procedure

_____ 1. How to consult with the District Attorney's Office.
_____ 2. How to secure a complaint.
   a. Necessity of having reports in DDA's box by 8:30 A.M.
   b. When Warrant Detail of DD will sign complaint.
_____ 3. How to serve a warrant.
_____ 4. How to respond to a subpoena.
_____ 5. How to prepare for court appearance.
   a. Original notes.
   b. Preparing evidence.
   c. Appearance and dress.
   d. Conduct in court.
   e. Court overtime.
   f. Fees not permitted.
_____ 6. How to testify in Municipal Court.
   a. Observe sessions of the Municipal Court.
   b. Trial by court.
   c. Jury trial.
   d. Traffic, criminal, drunk driving.

_____ 7. How to testify in Superior Court.
_____ 8. How to testify in Juvenile Court.
_____ 9. How to testify at an insanity hearing.
_____ 10. How to testify in Civil Court.
_____ 11. How to serve a subpoena.

Property

_____ 1. How to handle lost and found property.
_____ 2. How to place property in property room.
_____ 3. How to impound automobile and release.
_____ 4. How to return property, property receipt.
_____ 5. How to describe property in detail. Search for numbers or identifying marks, and determine in what manner an article differs from all others.
_____ 6. How to establish ownership of property.
_____ 7. How to investigate case of abandoned bicycle; checking files after locating factory and police numbers.
Checking with other local departments.
_____ 8. Property receipts—PR 201; 332-335.
_____ 9. 2973—N.S.
_____ 10. Disposition of property in possession of Department.
   a. Use of property disposition form.
_____ 11. Procedure to follow upon receipt of evidence disposition card from Property Clerk. Reasons for prompt action.

Reports

_____ 1. How to prepare narrative and form reports.
   a. Heading.
   b. Synopsis—first paragraph, who, what, when, where, how.
   c. Body of the narrative report.
   d. How to write a report setting a follow-up.
   e. Recontacting complainants.
   f. How and when to write a suspending report.
   g. How to write a closing report.
_____ 2. How to prepare a regular M.O. crime report.
_____ 3. How to prepare a DMV accident form.
_____ 4. How to prepare an injury accident investigation form.
_____ 5. How to prepare a noninjury accident form.
_____ 6. How to prepare miscellaneous public report forms for accidents occurring on Eastshore Highway.
_____ 7. How to make a drunk driver report.
   a. Necessity of including driver's license information.

FIELD TRAINING GUIDE 299

        b. Necessity of sending teletype to DMV for complete record of convictions to be forwarded to DDA office.
        c. Last paragraph of report must indicate disposition of auto.
        d. G.O. A-53.4.
_____ 8. How to fill out a stolen and recovered auto form.
        Verifying APB radio and teletype bulletins.
        Procedure for cancelling APB's.
_____ 9. How to prepare a bicycle theft report.
_____ 10. How to make out a property receipt.
_____ 11. How to prepare a miscellaneous service complaint form.
_____ 12. How to prepare a message for the teletype and where to leave it. Refer to G.O. T-1.2.
_____ 13. How to make request for assistance of another beat officer or platoon.
        a. Requires authorization of sergeant.
_____ 14. How to request reassignment of a case.
_____ 15. How many carbon copies of above reports—routine.
General Rules
        a. 2 CC to JD on all cases involving juveniles, female victims 5050, unfit homes, arrested females under 18 years of age.
        b. CC to Captain of Service Division in cases where U.C. students are responsible for offenses.
        c. CC to beat officer in every case handled for that officer.
        d. CC to DD on all felonies, sex offenses, suicide attempts, and dead bodies.
        e. CC to S.I.B. and security detail in all cases involving matters of their concern.
        f. CC to Armed Services Police of accidents involving military personnel.
        g. CC to City Attorney and City Manager in all cases involving matters of their concern.
        h. CC to Personnel Captain on all cases of injury or exposure of officer on duty.
        i. CC (2 copies) to DDA in all cases where an arrest made or person advised to see him. On all CC to DDA reports where subject in custody, the name of the defendant will be typed at the top of the report.
        j. CC to Court in some cases.
        k. CC to Tax and License Administration when city property damaged.

Appendix J

California Commission on Peace Officers' Standards and Training Specifications

The Supervisory Course

Purpose

3-1. Specifications of the Supervisory Course: This Commission Procedure implements that portion of the Minimum Standards for Training established in Section 1005 (b) of the Regulations which relate to Supervisory Training.

Content and Minimum Hours

3-2. Supervisory Course Subjects and Minimum Hours: The Supervisory Course is a minimum of 80 classroom hours and consists of the following subjects:
    a. Introduction and Scope of the Course
    b. Duties and Responsibilities of the Police Supervisor
    c. The Supervisor's Relationship to Police Management
    d. Communication Principles
    e. Handling and Prevention of Complaints
    f. Motivating Employees to Work
    g. Leadership
    h. Psychological Aspects of Supervision
    i. Morale and Discipline

COURSES AND PROGRAMS

j. Performance Appraisal and Rating Procedures
k. Supervisory Decision Making
l. Making Duty Assignments
m. The Supervisory Training Function
n. How People Learn
o. Job Analysis
p. The Four Steps of Teaching
q. Lesson Plans
r. Instructional Aids
s. Roll Call Training
t. Practical Application
u. Evaluation of Instruction
v. Written Examinations

Professional Certification Program

Purpose

1-1. The Professional Certification Program: This Commission Procedure implements the Professional Certification Program established in Section 1011 (c) of the Regulations.

General Provisions

1-2. Eligibility: To be eligible for the award of a certificate, an applicant must be:
  a. A full-time, paid peace officer member of a California city police department, a California county Sheriff's department, the California Highway Patrol, the University of California Police, or the California State University and Colleges Police, OR
  b. A former full-time, paid peace officer member of a California city police department, a California county sheriff's department, or the California Highway Patrol, who at the time of application is serving as a full-time, paid peace officer as defined by California law.

1-3. Application Requirements:
  a. All applications for award of certificates covered in this specification shall be completed on the prescribed Commission form entitled "Application for Award of Certificate."
  b. Each applicant shall attest that he or she subscribes to the Law Enforcement Code of Ethics.
  c. The application for a certificate shall provide for the following recommendation of the department head:

(1) "It is recommended that the certificate be awarded. I certify that the applicant has complied with the minimum standards set forth in Section 1002(a)(1), (2), and (4) of the Commission's Regulations, is of good moral character, and is worthy of the award. My opinion is based upon personal knowledge or inquiry, and the personnel records of this jurisdiction substantiate this recommendation."

(2) When a department head is the applicant, the above recommendation shall be made by the department head's appointing authority such as the city manager or mayor. Elected department heads are authorized to submit an application for approval by the Commission.

Education, Training, Experience

1-4. Basis for Qualification: To qualify for award of certificates, applicants shall have completed combinations of education, training, and experience as prescribed by the Commission.
  a. Education Points: One semester unit shall equal one education point and one quarter unit shall equal two-thirds of a point.
  b. Training Points: Twenty classroom hours of police training approved by the Commission shall equal one training point.
  c. When college credit is awarded for police training, it may be counted for either training points or education points, whichever is to the advantage of the applicant.
  d. Law enforcement experience in California as a full-time, paid peace officer member of a city police department, a county sheriff's department, or the California Highway Patrol may be acceptable for the full period of experience in these agencies.
  e. In other law enforcement categories designated by the Commission, the acceptability of the required experience shall be determined by the Commission, not to exceed a maximum total of 5 years.

Professional Certificates

1-5. The Basic Certificate: In addition to the requirements set forth in paragraphs 1-2 and 1-3, the following are required for the award of the Basic Certificate:
  a. Shall have completed the probationary period prescribed by the employing jurisdiction, but in no case of less than one year.
  b. Shall have satisfactorily completed the P.O.S.T. Basic Course or its equivalent as determined by the Commission.

COURSES AND PROGRAMS

1-6. The Intermediate Certificate: In addition to the requirements set forth in paragraphs 1-2 and 1-3, all of the following are required for the award of the Intermediate Certificate:
   a. Shall possess or be eligible to possess a Basic Certificate.
   b. Shall have acquired the following combinations of education and training points combined with the prescribed years of law enforcement experience, or the college degree designated combined with the prescribed years of law enforcement experience.

Minimum Training Points Including POST Basic Course	15	30	45	POST Basic Course	POST Basic Course
Minimum Education Points	15	30	45	Associate Degree	Baccalaureate Degree
Years of Law Enforcement Experience	8	6	4	4	2

Table 4.

1-7. The Advanced Certificate: In addition to the requirements set forth in paragraphs 1-2 and 1-3, the following are required for the award of the Advanced Certificate:
   a. Shall possess or be eligible to possess the Intermediate Certificate.
   b. Shall have acquired the following combinations of education and training points combined with the prescribed years of law enforcement experience, or the college degree designated combined with the prescribed years of law enforcement experience.

Minimum Training Points Including POST Basic Course	30	45	POST Basic Course	POST Basic Course	POST Basic Course
Minimum Education Points	30	45	Associate Degree	Baccalaureate Degree	Master's Degree
Years of Law Enforcement Experience	12	9	9	6	4

Table 5.

1-8. The Management Certificate: In addition to the requirements set forth in paragraphs 1-2 and 1-3, the following are required for the award of the Management Certificate:

    a. Shall possess or be eligible to possess the Advanced Certificate.

    b. Shall have been awarded a baccalaureate degree or an associate degree or not less than 60 college semester units at an accredited college as defined in Section 1001 (a) of the Regulations.

    c. Shall have completed satisfactorily the Middle Management Course or its equivalent as provided in Section 1008 of the Regulations.

    d. For a period of two years shall have served satisfactorily as a department head, assistant department head, or as a middle manager as defined in Sections 1001 (h), (c) and (l) of the Regulations. The required experience shall have been acquired within five years prior to date of application.

    e. The Management Certificate shall include the applicant's name, official title, and name of jurisdiction. When a holder of a Management Certificate transfers as an assistant department head or middle manager to another jurisdiction and upon the completion of one year of satisfactory service in a new department, upon request, a new certificate may be issued displaying the name of the new jurisdiction.

1-9. The Executive Certificate: In addition to the requirements set forth in paragraphs 1-2 and 1-3, the following are required for the award of the Executive Certificate:

    a. Shall possess or be eligible to possess the Advanced Certificate.

    b. Shall have been awarded a baccalaureate or associate degree or higher, or no less than 60 college semester units at an accredited college as defined in Section 1001 (a) of the Regulations.

    c. Shall have completed satisfactorily the Executive Development Course or its equivalent as provided in Section 1008 of the Regulations.

    d. For a period of two years shall have served satisfactorily as a department head as defined in Section 1001 (h) of the Regulations. The required experience shall have been acquired within five years prior to date of application.

    e. The Executive Certificate shall include the applicant's name, official title, and name of jurisdiction. When a holder of an Executive Certificate transfers as a department head to another jurisdiction and upon the completion of one year of satisfactory service in a new department, upon request, a new certificate may be issued displaying the name of the new jurisdiction.

Middle Management Course

Purpose

4-1. Specifications for Middle Management Course: This Commission Procedure implements that portion of the Minimum Standards for Training established in Section 1005 (c) of the Regulations which relate to Middle Management Training.

Content and Minimum Hours

4-2. Middle Management Course and Minimum Hours: The Middle Management Course is a minimum of 100 classroom hours and consists of the following subjects:

4-3. Introduction:
    a. Course Orientation
    b. Role of Police in Society

4-4. Organization and Management:
    a. Principles of Administration
    b. Modern Police Organization
        1. Line Functions
        2. Administrative Functions
        3. Auxiliary Functions
    c. Role of Middle Manager

4-5. Motivation:
    a. Human Relations in Management
    b. Techniques of Supervision
    c. Psychology of Leadership
    d. Effective Communication
    e. Conference Leadership

4-6. Implementation:
    a. Research, Planning, and Analysis
    b. Deployment and Utilization of Personnel
    c. Financial Planning, Execution, and Control
    d. Community Relations Program Management
    e. Information Management
    f. Training Program Management
    g. Personnel Management
    h. Planning for the Future

4-7. Individual Projects: Course administrators may require each trainee to complete a study project related to one or more of the subjects in the Middle Management Course including a written report of the project, including findings and conclusions.

Basic Course

Purpose

1-1. Specifications of Basic Course: This Commission Procedure implements that portion of the Minimum Standards for Training established in Section 1005 (a) of the Regulations which relate to Basic Training.

Content and Minimum Hours

1-2. Basic Course Subjects and Minimum Hours: The Basic Course is a minimum of 200 hours and consists of the following subjects and minimum number of hours of instruction.

1-3. Introduction to Law Enforcement: 10 hours
    a. Criminal Justice System.
    b. Ethics and Professionalization
    c. Orientation

1-4. Criminal Law: 16 hours
    a. Criminal Law (Penal Code)
    b. Laws of Arrest

1-5 Criminal Evidence: 8 hours
    a. Rules of Evidence (Evidence Code)
    b. Search and Seizure

1-6 Administration of Justice: 4 hours
    a. Court System
    b. Courtroom Demeanor and Testifying

1-7. Criminal Investigation: 34 hours
    a. Assault Cases
    b. Auto Theft Cases
    c. Burglary Cases
    d. Collection, Identification, and Preservation of Evidence
    e. Crime Scene Recording
    f. Injury and Death Cases

g. Interviews and Interrogations
   h. Narcotics and Dangerous Drugs
   i. Preliminary Investigation
   j. Robbery Cases
   k. Sex Crimes
   l. Theft Cases

1-8. Community-Police Relations: 20 hours
   a. Discretionary Decision Making
   b. General Public Relations
   c. Human Relations
   d. Local Programs
   e. News Media Relations
   f. Race and Ethnic Group Relations
   g. Role of Police in Society
   h. Role-Playing Demonstration

1-9. Patrol Procedures: 40 hours
   a. Alcoholic Beverage Control Laws
   b. Crowd Control
   c. Disaster Training
   d. Disorderly Conduct and Disturbance Cases
   e. Domestic and Civil Disputes
   f. Field Notetaking
   g. Intoxication Cases
   h. Mental Illness Cases
   i. Missing Persons
   j. Patrol and Observation
   k. Report Writing
   l. Tactics for Crimes in Progress
   m. Telecommunications

1-10 Traffic Control: 20 hours
   a. Citation: Mechanics and Psychology
   b. Driver Training
   c. Drunk Driving Cases
   d. Traffic Accident Investigation
   e. Traffic Directing
   f. Traffic Laws (Vehicle Code)
   g. Vehicle Pullovers

1-11. Juvenile Procedures: 8 hours
   a. Juvenile Laws
   b. Juvenile Procedures

1-12. Defensive Tactics: 14 hours
   a. Arrest and Control Techniques
   b. Defensive Tactics
   c. Transportation of Prisoners and the Mentally Ill

1-13. Firarms: 12 hours
   a. Legal Aspects and Policy
   b. Range
   c. Special Weapons

1-14. First Aid: 10 hours

1-15. Examinations: 4 hours

1-16. Total Required Hours: 200 hours

Advanced Officer Course

Purpose

2-1. Specification of Advanced Officer Course: This Commission Procedure implements that portion of the Minimum Standards for Training established in Section 1005 (d) of the Regulations which relate to Advanced Officer Training.

Course Objective

2.2 Advanced Officer Course Objectives: The Advanced Officer Course is designed to provide updating and refresher training for law enforcement officers. Flexibility is to be permitted in course content and manner of course offering in order to meet changing conditions and local needs, and yet remain consistent with the updating-refresher concept.

Course Content

2-3. Advanced Officer Course Content: Required General Updating-Refresher Subject Matter: The content shall devote no less than 10 hours to any combination of the following subjects:
   New Laws
   Recent Court Decisions and/or Search and Seizure Refresher
   Officer Survival Techniques
   New Concepts, Procedures, Technology
   Discretionary Decision Making (Practical Field Problems)

Elective Subject Matter: The course may contain such other currently needed subject matter which fall within the topical areas of the Basic Course Commission Procedure D-1. It is suggested elective subjects focus on current and local problems or needs of a general, rather than specific, nature.

2-4. Presentation and Curriculum Design: Curriculum design and the manner in which the Advanced Officer Course is proposed to be presented may be developed by the advisory committee of each agency certified to present the Advanced Officer Course and shall be presented to the Commission for approval.

2-5. Minimum Hours: The Advanced Officer Course shall consist of a minimum of 20 hours.

Index

A
Absenteeism, 93
Acceptance, of communication, 157
Accidents:
    responsibility for, 87–88
    waste resulting from, 35
Accountability, of decision-maker, 55
Action:
    and creation of dissatisfaction, 180–81
    of employees, 180
    and grievances, 180–81, 183–84
    political, 120
    as source of complaints, 179
Activity:
    definition of, 108
    unnecessary, 74
Activity-Interaction-Sentiment Model, 107
Adequacy, of communication, 156
Adjustment, 99–100
Administration, 2, 31, 32
    philosophy of, 7
    of training, 230
Advanced Officer Course 308–309
Africa, and unions, 113
Alcohol, and role of supervisor, 58–60

Alioto, Joseph, 112
American Bar Association, 224
American Federation of State, County, and Municipal Employees, 112, 117
Anger, control of, 76–77
Appeals:
    disciplinary, 132–34
    hearing, 136
    to higher motives, 128
Appearance:
    personal, 79, 138
    teacher's, 248
Appraisals, performance:
    descriptive scale, 214
    evaluation, 213
    good system, elements of, 218
    improvement of, 219
    and incidental behavior, 215
    and incompetence, 215
    and knowledge of employee, 216
    leniency in, 216
    and measurement, 217
    and negligence, 215
    numerical scale, 214
    and objectives, 243
    and potential value, 217

312 INDEX

Appraisals, performance, continued
   and prejudice, 215
   problems and errors in, 215
   ranking system, 214
   severity in, 216
Arbitration:
   compulsory, 119–20
   voluntary, 119
Areas, of inquiry, 44
Assessment centers, 46–47
Assistance, of subordinates, 82
Associations, 92, 103–122
   activities of, 120–22
   police, 112–17
   types of, 117
   See also Groups
Atmosphere, work, 138, 233
Attitude, 28
   businesslike, 78
   cheerful, 84
   decision making, 196
   employee, 180
   friendly, 84
   as source of complaints, 179
   supervisor's, 180
Audio-visual teaching aids, 252
Authority:
   and approval before taking action, 30
   areas of, 30–31
   delegation of, 30
   full, 30
   limited, 30
   lines of, 3
   system, 111
Availability of supervisor, 85

B

Baccalaureate degree, 4, 229
Bargaining, collective, 113
Basic Course (police training), 306–308
Behavior, 50, 51
   incidental, 215
Belonging and love needs, 96
Berkeley, California, Police Department:
   Field Training Guide, 285–99
   training manual, 234
   Vollmer plan, 10
Blame, assignment of, 68
Blue flu, 122
Boards, civilian review, 117
Booking, of accused officers, 145
Bopp, William J., 117
Boss, supervisor as, 4, 5, 40
Boston, Massachusetts, police strike in, 120
Bristow, Allen, 47, 190, 242
Budget, 33
   and waste control, 34
Bulletin, training, 239
Butler, John, 158

C

California:
   code of ethics, 126–27
   colleges offering criminal justice and police education, 230
   form for rating supervisors, 278–79
   questionnaire for supervisors, 280–84
   supervisors' rating scale, 275–77
   training programs, 4
   training standards, 228–29
California Peace Officers Association, code of ethics, 126–27
California Program for Peace Officers Training, 275–84
California State Universities, 240
California survey of supervisory training, 47
Candidness, importance of, 81
Centers, assessment, 46
Central tendency, error of, 216
Certification, professional program of, 301–304
Change:
   and confusion, 76
   economic, 3, 7
   and management, 17, 18
   political, 3
   social, 3, 7
Charts:
   organization, 25, 29, 111
   training, 252
Cheerfulness, and supervisor, 84
Chicago Police Department,

INDEX 313

organization chart of, 25
Clarity:
    in communication, 151
    in decision making, 192
Classical model, 11
Climate:
    of organization, 5
    -setter, 54
Clinical approach, 140
Code of ethics, 126–27
Codification, of laws and regulations, 8
Colleague model, 11, 12
Collective bargaining, 113
College, 230
    college-level programs, 240
    external degree programs, 230
    growth of criminal justice education, 229
    requirements for promotion, 47
Command, chain of, 163
Commission on Law Enforcement and Administration of Justice, 4, 181
Commission on Peace Officers' Standards and Training. *See* Peace Officers' Standards and Training
Communication, 19, 148–66
    acceptance of, 157
    adequacy of, 155
    barriers to, 161
    categories of information, 156
    channels of, 83–84
    clarity of, 151
    consistency of, 153
    continuity of, 153
    distribution of, 155
    follow-up on, 160
    group discussion, 157
    horizontal, 150
    and human element, 151
    interest level of, 157
    lateral, 150
    and layers of insulation, 163
    learning to listen, 162
    positional, 149–50
    principles of, 151
    with subordinates, 81
    and supervisor's role, 55
    timeliness of, 155
    timing of, 155
    uniformity of, 157
    verbal, 165
    vertical, 149
    written, 165
Community, 7
    cultural influence of, 53
    growing urban, 2
Compensation, grievances about, 181
Competency:
    instructor's, 232
    professional, 78
Competition, as incentive, 98
Complaints, 83, 177–86
    choosing investigator for, 142
    and decision making, 184
    definition of, 178
    determination of jurisdiction for, 183
    disposition of, 145
    dissatisfaction, sensitivity to, 181–82
    evaluation of, 143
    final disposition of, 146–147
    follow-up on, 184
    forestalling, 182
    form, 260
    handling personnel with, 140–47, 183
    interviews, 143, 174–75, 183
    investigation of, 141, 183
    investigation reports on, 146
    investigation techniques for, 144–45
    about misconduct, 256–59
    receiving, 142
    sources of, 178–79
Conciliation, supervisor's role in, 54–55
Conduct:
    misconduct, 256–59
    standards for, 74, 75
Conference method of instruction, 241–42
Confidence, 67
    appearance of, 83
    between leader and group, 110

Confidence, continued
  respecting, 84
  winning of, 80–84
Conformity, 107
Consideration, importance of, 88–89
Consistency, importance of, 75, 153
Contingency planning, 195
Continuity:
  in communication, 153
  in teaching, 250
Control, 16, 26, 27, 28, 31
  and discipline, 128
  self-control, 76
  span of, 20
  of subordinates, 64
Cooperation, 2, 20
Coordination, 21, 24, 32
  of individuals and groups, 3
Cost of living, 113
Counseling, 168–76
  of subordinates, 3
  supervisor's role in, 58
  See also Interviews
Courses:
  advanced officer, 308–309
  basic, 306–308
  middle management, 305–306
  new supervisor, 47
  periodic, 239
  professional certification, 301–304
  supervisory, 300–301
Courtesy, importance of, 88–89
Credit, 77
Crime prevention, 3
Crime rate, 1,
Criticism, 66, 74
  of supervisors, 81
Culture:
  definition of, 53
  influence of, 53
  leadership in context of, 53–54
Culver City, California, training manual, 234

D

Decision-making: An Annotated Bibliography, 189

Decision making, 7, 22, 185, 188–202
  accountability for, 55
  and alternatives, 193
  and clarification of problem, 192
  definition of, 192
  elements of, 189
  errors in, 55
  factors influencing, 195
  and gathering facts, 192
  importance of, 188
  and indecision, 190
  outline for, 191
  and planning, 197
  preparation for, 191
  and promptness, 78–79
  public nature of, 190
  and selection of decision maker, 191
  supervisor's role in, 55
  and timing, 55
  See also Planning
Decisions:
  explanation of, 196
  implementation of, 196–97
Delegation, of authority, 32
Delinquency, juvenile, 1
Descriptive scale, 214
Development:
  capacity for, 4
  supervisor's, 2, 3
Diction, teacher's, 248
Differences, human, 50–51
Direction, 21, 32
Discipline, 125–47, 194
  building a case, 134–37
  characteristics of poor, 126
  and clinical approach, 140
  and code of ethics, 126
  and control, 128
  definition of, 125
  desciplinary interview, 174
  and dismissal, 60
  evaluation of offense, 130–31
  and hearings, 132
  and morale, 94
  nature of, 127
  negative, 127–28

positive, 127
preventive, 140
and problem drinkers, 59
securing, 137
and supervisor's role, 56
and suspension, 59
and teacher, 249
and transfers, 131
See also Complaints
Dismissal, 60, 130
Disobedience, as technique of labor, 121
Disposition, importance of good, 248
Dissatisfaction:
  definition of, 178
  recognizing, 181–82
Distribution, of communication, 155–56
Division charts, 29
Drinker, problems of, 58–60
Drucker, Peter, 190
Duties, 28
  administrative, 32
  supervisory, 31–34
  training, 33

E
Education, 229
  college-level, 48, 230, 238, 240
  of supervisors, 47
  and supervisor's responsibilities, 230
Effectiveness, of supervision, 5
Efficiency:
  level of, 5
  and morale, 94
Ego-ideal, supervisor's role as, 55
Eisenhower, Dwight D., 71
Emerson, Ralph Waldo, 75
Emotion, display of in interview, 172
Empathy, importance of, 85
Employee associations, 92, 103–122
  activities of, 120–22
  police 112–17
  types of, 117
Employment:
  outlook, 7

permanent status, 264–69
England, and unions, 113
Enthusiasm, importance of teacher's, 248
Environment, 52, 117
  law enforcement, 7–15
Equal Rights Amendment, 222
Equipment, waste of, 35–36
Errors, responsibility for, 77
Esteem needs, 96
Evaluation, 7
  of performance, 224
  personnel evaluation form, 261–63
Examinations (tests), validity and reliability of, 42
  written, 42
Example, importance of setting a good, 139
Exception principle, 154
Exchange theory, 108
External degree programs, 230

F
Fact finding, 119
Factor analysis, 43
Facts:
  fact-finding, 192
  gathering of, 192
  and objectivity, 193
  sources of, 192
  See also Investigation
Failure, reasons for, 3
Favoritism, 89–90
Fear, 68
  and discipline, 128
Federal Bureau of Investigation, National Police Academy, 240
Federal Republic of Germany, and unions, 113
Feedback, 194, 197
  public, 54
Field instruction, 234
Field Instructor's Schedule, 235
Field Training Guide, 235
  Berkeley, 285–99
Film Guide, Police, 242
Finance, in decision making, 194

Firearms, disobedience of limitations on use of, 121
Firmness, importance of, 139
Follow-through, 73, 211
Follow-up, in communication, 160
Ford Motor Company, employee policies, 4
Formal (traditional) organization, 11
Forms:
    complaint, 256-60
    personnel evaluation, 261-63
    self-analysis, 280-84
    supervisor's rating scale, 275-77
    supervisory rating, 278-79
France, and unions, 113
Fraternal Order of Police, 112, 116
Friction, between individuals and groups, 3
Friendliness, importance of, 84
Friendship, with subordinates, 89-90
Frustration, effect of on job, 5
Function charts, 29

G
Gabard, E. C., 190
Gammage, Allen, 241-42
Goals, modification of, 107
Golembiewski, Robert T., 11-12
Gourley, G. Douglas, 134, 163
Grievances, 31, 83, 94, 114, 177-86
    attitudes, actions, and complaints, 179-81
    and compensation, 181
    definition of, 177
    dissatisfaction, sources of, 178-81
    handling of, 183-86
    and interviews, 183
    personal problems, 178
    plan of action, 183
    recognition of, 181
    and work assignments, 178-79
Groups
    Activity-Interaction-Sentiment Model, 107-108
    American police unions, 114-17
    analysis and management of, 107-112
    discussion, 157

effective work, 110
effects of on individuals, 106
employee, 103-122
formal and informal, 109
friction between individuals and, 3
functions of, 105
and labor relations, 118
and morale, 92-93
police employee, 112-14
police groups today, 115-17
and police militancy, 117-18
typology, and characteristics of, 104-105
    formal, 104
    informal, 104
    semiformal, 105
See also Organization, Employee associations
Guilds. See Associations, Unions

H
Habits:
    teachers', 248
    work, influence of culture on, 53
Hatred, in subordinates, 61
Health, of subordinates, 86
Hearings:
    appeal, 136
    building a case, 134-37
    disciplinary, 132-34
Heat, waste of, 36
Heredity, 52
Hero worship, 55
History, of law enforcement, 8
Homans, George C., 107-108
Honesty, importance of, 80
Human differences, 50-53
Human element, in communication, 151
Human relations:
    approach, 4-5, 11
    and informal organization, 11

I
Identification, with others, 66
Impartiality, importance of, 89-90, 139
Incentives, 98-99

Incompetence, 215
  of rater, 215–17
Indecision:
  avoiding, 194
  in decision making, 190
Indiana University, 230
Individual:
  differences, 50–53
  involvement in decision making, 193
  qualities of, 66
  recognition of subordinate as, 85–86
Induction interview, 172, 235
Industrial revolution, 114
Informal organizations, 11
Information:
  categories of, 156
  imparting, 241
  about job, 236
  and objectivity, 193
  for subordinates, 81
  about work, 237
Innovation, as function of managers, 22
Inquiry, areas of, 44
Inspections, importance of, 137–38
Instruction:
  application of, 242
  audio-visual teaching aids, use of, 252
  characteristics of good teacher, 244–49
  and conferences 241–42
  and context, 249
  continuity, importance of, 250
  course work, 241
  and discipline, 249
  experts as instructors, 250
  field, 234
  imparting information, 241
  introduction to, 242
  learning process, speeding up, 250–51
  lesson planning, 243–44, 245–47
  methods of, 241–42
  and needs of student, 249
  organized, 241
  presentation of, 242
  principles of, 244
  reinforcement, of, 250
  steps in, 242–43
  and tests, 243
Insulation effect, 163
Integrity, importance of, 79
Interactions, 108
Interest:
  in communication, 157
  special, 193
  in subordinates, 86
  of subordinates, 82
International Association of Chiefs of Police, 240
  and assessment centers, 46
  as source of audio-visual teaching aids, 252
International Brotherhood of Police Officers, 117
International Brotherhood of Teamsters, Chauffeurs, Warehousemen, and Helpers, 112, 117
International City Managers' Association, 9, 22
International Conference of Police Associations, 116
International Police Unions, 113
Interviews, 168–76
  with complainant and witnesses, 143–44
  complaint, 173, 183
  disciplinary, 174–75
  elements of, 169
  emotional involvement during, 172
  establishment of rapport during, 170
  grievance, 173
  induction, 172, 235
  nondirective, 169
  personal problem, 174
  and privacy, 170
  problem-solving, 171
  progress, 172
  special, 172
  termination, 175
Investigation, 192

Investigation, continued
  of complaint, 141, 145
  file, preparation of, 135
  reports, 146
  techniques, 144
  See also Facts
Item analysis, 43
Ithaca, New York, strike at, 120

J
Japan, and unions, 113
Job analysis, 29
Job conditions, 97
Job information, 236–37
Job planning, 207–208
  analysis, 208
  effects of, 207
  steps in, 207
John Jay College of Criminal Justice, 230
Justice Research Associates, Costa Mesa, California, 46
Juvenile delinquency, 1

K
Kenney, John P., 8, 9, 12
Knowledge:
  inadequacy of, in rating, 216
  of supervisor, 73, 78

L
Labor movement, 114
Labor relations, 118–22
  negotiation techniques, 119
  and supervisor, 122
Law, 193
  enforcement of, 3, 7
  historical and contemporary, 8
Law and order issue, 117
Law Enforcement Assistance Association, 230
Laws, 8
Leadership, 71–90, 275
  in context of culture, 53–54
  improvement of ability, 72–80
  in interview, 171
  roles, 54–58
  rules, 72
  and winning confidence, 80
  and winning loyalty, 84
  and winning respect, 72
Learning process, methods of speeding up, 250–51
Leniency, in performance appraisal, 216
Lesson:
  planning, 243–47
  plans, 245–47
Levine, Jacob, 158
Lewin, Kurt, 158
Likert, Rensis, 110
Line and staff organization, 24–27
Listening:
  in counseling and interviewing, 168–69
  learning art of, 162–63
Litigation, for employee benefits, 121
Lobbying, by police organizations, 120
London Police Department, 53
Love and belonging needs, 96
Loyalty, winning of subordinates', 84

M
Mager, Robert F., 243
Maintenance of order, 9
Maladjustment, 100
Management:
  functions of, 19, 37
  process, 18
  and supervision, 29
Management by Objectives, 12–15
Management by Results, 12–15
Manual:
  departmental, 5, 29
  training, 4
Maslow, Abram H., 96
Measurement, in performance appraisal, 217
Mediation, 199
Mental illness, 60–61
Methods:
  of instruction, 241–42
  of planning work assignments, 210
  of waste control, 34
  work, 30
Michigan State University, National Institute on Police and Community Relations, 230

INDEX

Middle management, course in, 305–306
Militancy, of police, 112, 117–18
Minority groups, 7, 57
Misconduct, complaints of, 256–59
Mistakes, of supervisor, 55
Misunderstandings, 7
Modified plan, 10
Money, 98
Mood, in decision making, 195
Morale, 3, 75, 92–102
   in decision making, 194
   definition of, 92–93
   factors in, 94
   and human wants and needs, 96
   importance of, 93
   level, determination of, 101
   and productivity, 93
   and public relations, 94
   and satisfaction of human needs, 98
   supervisor's responsibility for, 95
Motivation, 19, 63–70
   and assignment of blame, 68
   checklist for, 69
   and delegation of responsibility, 65
   failure to provide, 67
   and influence of culture, 53
   for knowledge, 238
   positive and negative, 68
   praise as, 75
   principles of, 63–70
   and recognition, 66
   and self-confidence, 67
   and self-involvement, 63
   of subordinates, 3
   and training, 63–64
Municipal Police Administration, 9, 22
Municipal Police Training Council, 229

N

National Association of Government Employees, 117
National Union of Police Officers, 117
Needs, individual:
   belonging and love, 96
   effects on groups, 108–112
   esteem, 5
   and groups, 105–106
   physiological, 96
   realization of, 96
   safety, 96
   satisfaction of, 98
   self-actualization or recognition, 96
   student, 249
Negligence, in rating, 215
Negotiation techniques, 119
Neurosis, 61
New York (state):
   supervisory training course, 270–74
   training programs, 4
   training standards, 228–29
New Zealand, and unions, 113–14
New Zealand Police Association, 113
New Zealand Police Officers Guild, 113
Northwestern University, 240
   Traffic Institute, 230
Numerical scale, for performance appraisal, 214

O

Oakland, California:
   complaint form, 260
   lobbying by police officers, 120
Objective setter, supervisor's role as, 54
Objectives, 137–39
   departmental, 3
   establishment of, 19
   performance, 243
   reasonable, 138
Oklahoma City police strike, 112
Orders, 3
   follow-up on, 160
   selling to subordinates, 158
   verbal, 165
   written, 165
Organization:
   charts of, 25, 29
   climate of, 5
   definition of, 22
   duties in, 31–32
   employee, 92, 103–122
   limits of, 23–24
   line and staff, 24–27
   line duties in, 24–27
   local, 115–16

Organization, continued
  models, 10
  national, 116
  need for, 23
  plan, 23-24
  policies of, 30
  principles of, 22-23, 34
  pyramid, 24
  semi-military, 10
  state, 116
  and structure, 9
Organizing Men and Power, 11, 12
Orientation training:
  field instruction, 234
  line supervisor's responsibilities in, 234
  and reception of recruits, 233-34
  recruit school, 234
Overprotectiveness, avoiding, 139
Overtime, 2

P
Pakistan, and unions, 113
Pasadena, California, training manual, 234
Patience, importance of, 76
Peace Officers' Commissions, as source of audio-visual teaching aids, 252
Peace Officers' Research Association of California, 112
  code of ethics, 126-27
Peace Officers' Standards and Training, 4, 48, 228, 252
  creation of, 228-29
  field training guide, 235
  function of, 228-29
  specific courses:
    Advanced Officer, 308-309
    Basic, 306-308
    Middle Management, 305-306
    Professional Certification 301-304
    Supervisory, 300-301
  supervisors' form for rating, 278-79
  supervisors' questionnaire, 280-84

  supervisors' rating scale, 275-77
Peer, 2, 128
  ratings, 45-46
  status in group, 66
  supervisor as, 56
Pensions, 2
Performance appraisal. *See* Appraisals, performance
Periodic courses, 239
Personal life, 5, 178
  and problem interview, 174
Personality:
  defects, 3
  handicaps, 2
  misfits, 2
Personnel, 26, 27, 28, 30, 32
  deployment of, 33, 35
  evaluation form for, 261-63
  indication of drinking problem in, 58
  planning, 205-206
  practices, 118
  use of, 205-206
  waste of, 35
Personnel Board, California, 44
Physical condition, 79
  and decision making, 195
Physiological needs, 96
Planning, 21, 22, 31, 32, 205-211
  analysis and study for, 208
  contingency, 195
  and decision making, 197
  effects of, 207
  lesson, 243-47
  for personnel, 205-206
  procedures, 207
  program, 7
  staff studies, 199
  steps in, 197
  for subordinates, 3
  time, 205-206
  training course, 236
  and waste control, 34
  of work, 205
  work assignments, 209
  See also Decision making
Police Administration, 9, 12

INDEX

Police Benevolent Association, 112
Police Chief, The, 240
Police officer, and public office, 121
Police service, definition of, 3
Police Training in the United States, 242
Policies, 1, 3, 5, 29, 30, 32, 33, 37, 95
  decision-making, 193
  organizing, 31
Politics, in decision making, 193
Popham, W. James, 243
Population patterns, 2
Positional communication system, 150
Potential:
  employee, 4
  human, 52
  in performance appraisal, 217
Praise, importance of, 66, 75, 98
Prejudice:
  definition of, 89
  of rater, 215
Preparing Instructional Objectives, 243
President's Commission on Law Enforcement and Administration of Justice, 4, 181
Press, 194
Pressure, 5, 7
  social, 1
Preventive discipline, 140
Privacy, 170-71
Problems, 2, 178
  definition and clarification of, 192
  and personal interviews, 174
  solving of, 191
Procedures, 1, 3, 5, 29-30, 32-33
  failure to adhere to, 35
  selection of, 39-49
Productivity, 93-94
Professionalism, 1
Progress:
  interview, 172
  rate of, 4
  training charts, 252
Promises, 81-82
  and discipline, 128
Promotions, 4, 27, 30, 48-49, 225

Promptness, importance of, 82
Protector, supervisor's role as, 54, 139
Psychology, 5
  and differences among people, 50
  schools of, 52
Psychosis, 61
Public image of police, 1, 190, 193-94
Public relations, 5, 33
  and morale, 94
Punishment, 99, 111, 128-32
  shortcomings of, 132
Pyramid organization, 24

Q
Qualifications:
  for positions, 3
  of supervisors, 39-41
  See also Requirements
"Qualifications Appraisal Boards", 43

R
Rankin, Paul, 162
Ranking:
  of job conditions, 97
  system, 214
Rapport, in interviews, 170
Ratings:
  form for rating supervisors, 278-79
  scale, supervisor's, 275-77
Ratings:
  peer, 45-46
  scale of, 46
  service, 45
Realization needs, 96
Recognition:
  as incentive, 98
  as motivation, 66
Records, 26-28, 35, 37
  See also Forms
Recruitment, 223
  standards of, 228
Reeves, Elton T., 105
Regulations, 3
Reinforcement, importance of, 250
Relations, public, 5, 31, 33, 94

Relationships:
  interpersonal, 5, 31, 33
  with subordinates, 4-5, 16
  of supervisors, 29, 31
Reliability, of tests, 42
Reports, 26, 28, 30, 32, 37
  investigation, 146
Reprimands, 99
Requirements:
  baccalaureate degree, 4, 229
  See also Qualifications
Resignations, 134
Respect, importance of, 72
Responsibilities, 7, 31, 36
  for accidents, 87-88
  administrative, 31-34
  budget, 34
  coordination, 32
  delegation of, 28, 65, 209
  directing, 32
  for errors, 37, 77-78
  job, 2
  for morale, 95
  organizing, 32
  principles of, 34
  of supervisors, 3, 29, 31, 33
  training, 33-34
  written statements of, 29
Responsiveness of supervisor, 5
Review boards, civilian, 117
Reward, 111
Riverside, California, assessment center, 46
Robinson, Karl R., 158
Roles:
  supervisor as clarifier of, 56
  of supervisors, 54-58
Roll call training, 239
Rules, 3, 5, 6, 111

S

Sacramento County Sheriff's Department, 48
Safety, 87
  and new officers, 88
Safety needs, 96
Salaries, 2, 113-14, 118
  increases in, 4
  supplemental compensation, 4
  See also Compensation
San Francisco, California:
  lobbying, 120
  strike, 112
Scale:
  descriptive, 214
  numerical, 214
Scandinavia, and unions, 113
Scott, Clifford L., 32
Security, 98
Selection, 2, 4, 35
  of personnel for training, 232
  of supervisors, 27, 39-49
Self-confidence, 67
Self-control, 76
Self-improvement, 63-64
Self-involvement, 63
Seniority, 45
Sentiments, definition of, 108
Service Employees International Union, 117
Service ratings, 45
Severity, in performance appraisals, 216
Silander, Fred, 189
Skills and techniques, 5-6
  poor use of, 35
  social, tests of, 44
  teaching, 5
Slowdowns and speedups, 121-22
Social environment, 7
  and unions, 114
Social pressure, 1
Sociometric test, 44
Special interests, in decision making, 193
Speech (free), 73
Sponsor, supervisor as, 57
Staffing, 32
Standards:
  California, 228
  establishment of written, 137
  failure to adhere to, 35
  New York, 229
  for training, 228
  work, 74-75

Status, 110-12
Strikes, 112, 120
Structuralist approach, 11
Student, needs of, 249-50
Studies, management/supervisory staff:
   evaluation of solutions, 202
   identification of problem, 200-201
   interpretation of data, 201
Studies in Group Decision, 158
Subordinates:
   counseling, 3
   as individuals, 85-86
   motivating, 3
   planning for, 3
   supervisors as, 56
   training, 3
Supervision, 5, 6, 16
   definition of, 33
   functions of, 19, 20
   importance of, 1
   and management, 29
   one-to-one, 4
   permissive, 4
   training course, 270-74
Supervisor:
   attitude of, 180
   authority of, definition, 30
   and complaints, 179
   criticism of, 81
   and decision making, 188, 191
   definition of, 3, 17, 33
   determination of training needs, 231
   development of potential, 47
   and discipline, negative uses of, 128-29
   duties and responsibilities, 29, 31-34
   education, 47-48
   form for rating, 278-79
   functions, 7
   and groups, 109-112
   and labor relations, 122
   leadership, 71
   and mentally ill officers, 60
   and molding of subordinates, 52
   and motivation of employees, 63-70
   and organization, 21
   and police management process, 18
   P.O.S.T. course for, 300-301
   and principles of responsibilities, 34
   and problem drinkers, 58
   qualifications of, 39-40
   questionnaire for rating, 280-84
   rank of, 3
   rating scale for, 275-77
   and relationship with subordinates, 4, 5, 52, 65
   requirements for, 47
   responsibilities of, 230
   responsibilities for maintaining morale, 95
   responsibilities of line supervisor, 234
   responsive, 5
   roles of, 54-58
   selection of, 39-41
   setting good example, 139
   as subordinate, 56
   tools of, 5
   and training function, 228-54
   and waste control, 34
Suspension, of problem drinker, 59
Switzerland, and unions, 113
Symbol, supervisor's function as, 54
Sympathy, importance of, 85
Systems:
   analysis, 191
   and descriptive scale, 214
   evaluation, 213
   and numerical scale, 214
   ranking, 214

T

Tact, importance of, 88-89
Teaching, 244-54
   characteristics of good teachers, 244-49
Team approach, 11
Tension, 5
Termination, of interview, 175
Tests:
   factor analysis, 43
   group oral interview, 43-44
   item analysis, 43

Tests, continued
  reliability of, 42
  of social skills, 44
  sociometric, 44
  and training, 243
  validity of, 42
  written, 42
Time, planning of, 205–206
Timeliness, of communication, 155
Timing:
  and decision making, 197
  use of, 205–206
Traditional plan, 10
Training, 2, 32, 33, 224, 228–54
  advanced, 238
  bulletin, 239
  calendar, 240
  in California, 4
  college-level, 47–48, 240
  conference method of, 241
  course work, 241
  definition of, 229
  discipline as, 125
  duties, 33–34
  field instruction, 234
  field training guide, 285–99
  imparting information in, 241
  and induction interview, 235–36
  instructional methods for, 241–42
  instructional steps, 242
  lesson planning, 243–44
  and motivation, 238
  needs, 231
  in New York, 4
  and organized instruction, 241
  orientation, 233–35
  periodic, 239
  personnel, 29
  potential supervisors', 47
  principles of teaching, 244
  programs, 4, 5, 47, 232
  progress charts, 252
  and recruits, 236
  responsibility, 231
  roll call, 239
  standards, 228, 229
  subordinates, 2, 3, 4
  supervisory, 47–49
  and supervisory function, 228–54
  survey, 47
  work information, 237
  See also Peace Officers' Standards and Training
Training by Objectives, 12–15
Transfers, 30
  disciplinary, 131
Tucson, Arizona, strike in, 112
Turnover, 2, 93

U
Uniformity, in communication, 157
Unions, 107, 113
  development of police, 114–15
  today, 115–17
  See also Groups, Organizations
United States:
  Constitution, 10, 118
  unions in, 113
University of California at Berkeley, School of Criminology, 230
University of Southern California, Delinquency Control Institute, 230, 240
Urban community, growth of, 2, 7
Uses of Instructional Objectives: A Personal Perspective, 243
Utilities, waste of heat, light, and power, 36

V
Vacations, 2
Validity, of tests, 42
Vollmer plan, 10

W
Wages. *See* Salaries, Compensation
Wasserman, Paul, 189
Waste control, 34, 35, 36
Watson, Goodwin B., 158
Whyte, William F., 107
Wichita, Kansas, 234
Wickersham Commission, 9
Wilson, O. W., 9, 140, 197

Women:
    entrance of into police service, 222
    performance evaluation of, 224
    present location in police service, 222
    previous location in police service, 221
    promotion and opportunity for, 225
    recruitment of, 223
    and special considerations, 225
    training of, 224
    Women in Policing, 222

Women's movement, 222
Work assignments, 4
    elements of, 209
    flow of, 111
    follow-through, 211
    making, 211
    plannint methods for, 209–210
    as source of complaints, 178
Work information, 237
Work performance appraisal, 5
    See also Appraisals, performance
Work standards, 74